D1218738

THE NOVELIST AND MAMMON

THE NOVELIST AND MAMMON

Literary Responses to the World of Commerce in the Nineteenth Century

NORMAN RUSSELL

CLARENDON PRESS · OXFORD
1986

Oxford University Press, Walton Street, Oxford OX2 6DP
Oxford New York Toronto
Delhi Bombay Calcutta Madras Karachi
Kuala Lumpur Singapore Hong Kong Tokyo
Nairobi Dar es Salaam Cape Town
Melbourne Auckland

and associated companies in
Beirut Berlin Ibadan Nicosia

Oxford is a trade mark of Oxford University Press

Published in the United States
by Oxford University Press, New York

British Library Cataloguing in Publication Data

Russell, Norman
The novelist and mammon: literary responses to the world of commerce in the 19th century
1. English fiction — 19th century — History and criticism
2. Commerce in Literature
I. Title
823'809355 PR830.C6/

ISBN 0-19-812851-7

Phototypeset by Dobbie Typesetting Service, Plymouth, Devon
Printed in Great Britain at
The University Press, Oxford
by David Stanford
Printer to the University

For my Mother,
and my brother
Harold

PREFACE

Since the publication, in 1903, of Louis Cazamian's *Le Roman Social en Angleterre*, there have appeared many studies of the nineteenth-century novel which have explored the socio-political implications of these works, emphasizing and illustrating the individualistic, humanitarian reactions of such authors as Dickens, Kingsley, and Mrs Gaskell to the arid tenets of Utilitarianism.

Such works have been of enormous value in increasing our knowledge of Victorian literature, but in the main they have tended to contrast the nineteenth-century novelist with his fellow-intellectuals Bentham, Malthus, James Mill, and the rest, and by implication have suggested that these figures were in some way linked to a cruel capitalist 'system' against which the novelists reacted.

Of recent years a number of works have appeared that have explored the relationship between the novelist and the institutions of the City, and one welcomes the growing recognition that even the most reformist of novelists was a child of his own time in his attitude to commercial institutions. One would mention in particular James M. Brown's brilliant study, *Dickens: Novelist in the Market-Place* (1982), which has broken fresh ground in this respect.

My own book, while duly acknowledging the importance of prevailing economic theories, examines the responses of Victorian novelists to the thriving commercial life of the City, the world of stockjobbers and brokers, financiers and insurance promoters, and discusses how far these authors really understood the workings of those commercial institutions that they chose to depict. The work of a number of novelists is placed firmly in the context of the history of commercial life in the nineteenth century, showing how individual writers were in part conditioned in their literary responses by prevailing business practice, speculative manias, frauds, failures, and by the careers of some of the most celebrated entrepreneurs of the age.

Throughout the period there was a certain ambivalence towards the creation of wealth, an activity which was seen both as a vital element in the total progress of the nation and as a potential corrupter of common morality. From this uneasy dichotomy arose the favourite

Victorian metaphor of Mammon, which thus fittingly appears in the title of this book, and is referred to frequently throughout.

Specific acknowledgement for help received in the clarification of certain details of fact is given in the Notes, which function not only to give references, but to provide the reader with extra detail that he may find interesting. It gives me great pleasure to thank the Goldsmiths' Librarian, and the staff of the University of London Library, for the invaluable help that they have given me over the last few years in locating and sending me books and copies of documents.

This book is based upon a thesis accepted for the Degree of Doctor of Philosophy in the University of London. I wish to record my deep gratitude to Dr Michael Slater, who supervised the thesis, and to Professor Philip Collins and Professor Angus Easson, who encouraged me to rewrite it in book form.

N. R.
Liverpool, 1985

CONTENTS

LIST OF ILLUSTRATIONS

All illustrations are reproduced by courtesy of The Mary Evans Picture Library.

INTRODUCTION

The City: its Works and Ways in the Nineteenth Century

I

This book examines some of the ways in which English novelists reacted to the development of capitalism and its institutions from the early twenties to the eighties of the nineteenth century. During this period many events occurred in the commercial world that could not fail to attract the attention of the novelist, and these events offered further attractions in the various flamboyant personalities who dazzled—and often blinded—both the City and the crowds of investors who flocked to it with ideas of making a fortune.

In our own time, the word 'capitalism', first coined by James Mill in 1821, has gathered around it a number of political connotations that virtually condition a reader who sees the word in the title of a book to expect either a socialist condemnation or a monetarist defence. It may be said at once that in this book 'capitalism' is used in an economic, not a political, sense. In broad terms it is taken to mean the financing of public and private undertakings, in an age of dramatically swift industrial and commercial expansion, by means of money accumulated by specialist entrepreneurs. These men, the bankers, brokers, insurers, bill-discounters, and the rest, obtained their capital from various sources. Manufacturers and merchants deposited floating capital when they could, and borrowed at interest when the need arose, and at all times throughout the period, new enterprises sought the modest savings of the general public to raise their capital in the form of stock, shares, and debentures. This is, of course, an over-simplification of the capitalist 'system', but it describes adequately those aspects of its operation that most interested the novelists and the majority of their readers. The missing cheque, the forged bond, the rascally banker, the merchant turned miser—these became the stuff of fiction; but we seek in vain for any literary pleas for a collectivist alternative to the lures of the Money Market.

The reasons for this lack of interest in alternatives to the money-based society of the day are to be found in the hierarchic structures of nineteenth-century society, which produced a desire in both the lower and middle classes to emulate those whom birth had placed at the top of the social order.

Throughout the century there was a conflict between land and capital, in economic terms part of a universal struggle for income, but rationalized by apologists on both sides as a conflict of ideals. The aristocratic ideal of the landed gentleman, the guardian of property and patronage, was countered by the capitalist ethic of 'sturdy independence', achieved through effort and ambition. Competition was the vital and vitalizing force which all could exploit, free of aristocratic constraint, and the self-interest of the individual would serve to promote the good of society as a whole. It was morally reassuring to be told by writers of the 'self-help' school that material success and riches were attainable by the poorest men, and that the humblest operative could aspire to the captaincy of industry.[1]

In some ways, though, the 'new system' was fundamentally modelled on the old, and it was an aristocratic principle enunciated as early as the 1740s by David Hume that the successful entrepreneur had the right to bequeath the fruits of his industry to his descendants.[2] The novelist Mrs Catherine Gore saw this with her usual perspicacity when she made one of her characters observe: 'In this country, the aristocracy of wealth is beginning to be nicely balanced against that of descent: and a few generations may give it the ascendancy.'[3]

It is with the 'aristocracy of wealth' that this book is mainly concerned, and we shall see how the emerging 'middling classes' aspired to be equated socially with the old landed interest. The majority of wealthy merchants, bankers, and others desired no change in the status quo: 'Society' was the beau-idéal, and, to gain full social acceptance, some foothold in the counties was both desirable and

[1] Thus Edmund Potter could write in 1856: 'Society will ever remain composed of classes. Some are born with fortune, more are born without any . . . but *all* may possess industry which is, after all, the starting-point and by far the most valuable power.' Edmund Potter, *A Picture of a Manufacturing District* (Manchester, 1856), pp. 54–5.

[2] D. Hume, *Essays Moral, Political and Literary* (1741–2), ed. T. H. Green and T. H. Grose (1875), ii, p. 189.

[3] Mrs Gore, *The Banker's Wife* (1843), iii, p. 199. Volume and page refs. to Mrs Gore's novels are to the first editions. The dates of volume publication are given after the short title.

1. Mrs Catherine Gore.

necessary. However, in a world of entrenched attitudes and snobbery, the wealthy City magnate who aspired to broad acres frequently met with a rude shock.

No Victorian novelist more closely explored, or more thoroughly understood, the clash of old and new, of squirearchy and City magnate, than did Mrs Gore. During the 1840s she produced a number of novels the chief interest of which centres upon this theme of social acceptance and rejection, and it is certain that her fictional representations closely echo the situation in real life.

Mrs Gore first developed the theme of what she termed 'the aristocratic tendency' in *The Banker's Wife*, a novel which cleverly illustrates upper middle-class aspirations to landed proprietorship, and the secret disdain in which men of her banker Richard Hamlyn's standing were held by the aristocracy and gentry. As well as keeping a palatial town house in Cavendish Square, Hamlyn has an estate in the country, Dean Park, where he is able to gratify his ambition to become a landed proprietor. He is depicted as a fastidious man, with a genuine horror of 'manufacturing innovation', concerned lest a neighbouring vacant property 'fall into the hands of some moneyed speculator'.

Despite these aristocratic sentiments, Hamlyn is only tolerated by his landed neighbours. Lord and Lady Vernon, the local magnates, do not let Hamlyn forget 'their length of pedigree', which counters his 'length of purse', and even their housekeeper, receiving a friend of Hamlyn's to view Vernon's stately home, is scornful of what she calls 'the intrusion of the upstart tribe of Hamlyn the banker'. Nevertheless, Hamlyn is fascinated by these titled people, feeling that simply making their acquaintance in some way adds to his stature.

Three years after the appearance of this novel Mrs Gore returned to the theme with greater force in *The Man of Capital, or, Old Families and New*.[4] In this novel she describes a county's reaction to the intrusion of a member of the 'manufacturing classes', Mr Mordaunt, a Lancashire cotton-spinner who, having made a fortune, purchases the rural estate of Deasmarsh. Mordaunt revivifies a neglected backwater by creating a model estate, effecting 'agricultural improvements', and causing a rise in labourers' wages. The reaction of the squire and his son at the Hall is one of mean and venomous

[4] This is the second of two stories in the three-volume work *Men of Capital* (1846); it occupies vols. ii and iii.

snobbery allied to impotent hatred. Squire Cromer sneers at the idea of 'a *gentleman* belonging to the manufacturing districts', and his idle and dissipated son wonders 'why we are to be infested by the order of people whom engine-chimneys worm up out of the mud'. In the end he and his son create a fantasy about Mordaunt, believing, quite unjustly, that 'evil intentions lurked at the bottom of all his beneficence'.

Mordaunt initially defers to the Cromers' rank and station, but the continued taunts and sneers of father and son ultimately cause him to give vent to an uncontrolled and bitter condemnation: the mask of conformity drops, and in his anger he looks beyond the Cromers' somewhat tenuous social position to what they really are:

> A race of insignificant people . . . without energy to distinguish themselves from the clods from which they derive their subsistence! A race which has clung to the soil of Deaswold like the oak-apple to its oak, — only to deteriorate the parent stock: — poor, without talent or spirit to enrich themselves; — obscure, without containing within themselves the germ of ennoblement . . . incapable of extending their ideas or even their desires to the advancement of the country or amelioration of the human race! (II Ch. 12, pp. 262–3.)

Here, in a time of stress, is betrayed the result of a species of mental anguish which must often have beset men of Mordaunt's rank.[5]

At the end of Mrs Gore's novel there is no reconciliation between the two men, and Cromer and his wife leave the district. Mrs Gore clearly believes that an enlightened parvenu could achieve much good in a run-down rural environment, and, while not applying a blanket condemnation to the aristocracy — which would have been both unfair and untrue to reality — she is merciless in her castigation of obscurantism and unreasoning social envy. There is a 'confusion of classes', and those involved in the confusion must seek a *modus vivendi* for the good of all.

If Mrs Gore thought that her novels would succeed in alleviating the conflicting interests of land and capital, she must have been disappointed.

[5] In Charles Lever's novel *The Bramleighs of Bishop's Folly* (1868), Colonel Bramleigh the banker, failing to establish himself as a country gentleman, is stricken down by an apoplexy from which he does not recover. His doctor's opinion as to its cause is revealing:

> 'I'd call it rather the result of some wounded sensibility . . . his tone, so far as I can fathom it, implies intense depression. After all, we must say he met much coldness here: the people did not visit him, there was no courtesy, no kindliness shown him: and though he seemed indifferent to it, who knows how he may have felt it?' (ch. 23.)

The gulf was a vast one, at bottom a war of economic interests, and, nearer the surface, a complex problem of snobbery and the mythology of caste. In a vertically constructed society, movement had to be ever upward, and even the affronted mill-owner or banker would continue to set his eyes stubbornly upon a coveted position above his station.

The tensions arising from this uneasy movement of classes will be found echoed in many other novels of the period, where they frequently provide the chief focus of interest. In Miss Mulock's *John Halifax, Gentleman* (1856), we are shown how a poor but noble-hearted boy comes to terms with his social position in a society where 'the rich ground the faces of the poor', and 'the poor hated, yet meanly succumbed to, the rich'. Rising from poverty to the position of affluent tradesman, Halifax becomes socially acceptable to the middle-class Jessops, a family of doctors and bankers, but these in their turn lose the desirable patronage of the aristocratic Brithwoods, who will not attend their soirées if Halifax is also a guest. Here Miss Mulock skilfully depicts how the professional classes have a foot in both camps, seeking instinctively to rise into a social rank above theirs, and at the same time feeling at one with the well-to-do tradesmen who share the same or similar aspirations.

The Brithwoods are unpleasant people, but Miss Mulock's scorn is concentrated on the Earl of Luxmore, a degenerate and profligate nobleman, who regards 'the people' as a flock of sheep, and himself as their natural shepherd. Dishonourable and self-seeking, he is in turn contrasted with an old Tory gentleman, Sir Ralph Oldtower. This character, like Trollope's Roger Carbury in *The Way We Live Now* (1875), represents the true aristocratic virtues, uncorrupted by self-seeking and snobbery, and he forms a telling contrast to the do-nothing squires, dissipated earls, and money-grubbing gentry who disfigure the pages of many a Victorian novel. There is no concerted attack on the institution of aristocracy: it was reform, not abolition, that the novelist desired.

John Halifax is a self-made man, but his cheek glows with 'honest gratification and a pardonable pride' when the old knight publicly acknowledges him. Halifax admires Sir Ralph's 'fine old Norman face'— he would have agreed with Tennyson about the virtues of kind hearts and simple faith, but Norman blood still holds its romantic appeal for him. This is, perhaps, not surprising, for even as a poor boy he had known that his father had described himself as 'Guy

Halifax, Gentleman', and though his lineage remains for ever unrevealed, the consciousness of social gentility is always there.[6]

In all these novels one finds varying degrees of sympathy for the bourgeois aspirant when faced with the cruel hauteur of fictional aristocrats. Such novels present us with a motley array of nobles and gentry, some kindly and upright, generous and charitable—these often of 'Norman blood'—others mean-minded, decadent, vicious, or positively wicked. One may accept that the novelist exaggerates reality for effect, but there can be little doubt that he genuinely reflects a true social dilemma of his time, and that contemporary readers knew and recognized that dilemma for what it was.[7]

The three major classes, aristocratic, capitalist, and proletarian—the two former insisting on their own definitions of 'gentility' and claiming a virtual monopoly of true morality[8]—were joined in the nineteenth century by a fourth estate. This was the new middle class, neither aristocrat nor entrepreneur, but people who earned their livings by the exercise of professional knowledge. Engineers, architects, lawyers, physicians, surgeons; all became aware of their separate status, and began to band themselves into institutes and societies to enhance and confirm their distinct social identity. To this amorphous grouping

[6] The same theme is explored with particular sensitivity in Charles Lever's *The Bramleighs of Bishop's Folly* (1868). Colonel Bramleigh's whole family are preoccupied with questions of social caste. His elder son, the aspiring Temple Bramleigh, shrewdly remarks that 'every mile that separates you from the capital diminishes the power of your money'. He nevertheless aspires to the diplomatic service. His brother Augustus, more firmly bourgeois, is determined to show that 'the newly-risen gentry . . . Lombard Street people', are capable of noble deeds and sacrifices 'from which the peerage would shrink'.

[7] For a strident defence of the middle class and condemnation of the aristocracy, see Isaac Tomkins, *Thoughts upon the Aristocracy of England* (11th edn., 1835). 'The middle, not the upper class, are the part of the nation which is entitled to command respect . . . How long are they likely to suffer a few persons of overgrown wealth . . . to usurp, and exclusively to hold, all consideration, all importance?' This militant tone is, however, rarely heard. For a contemptuous treatment of the bourgeoisie, see in particular Bulwer's *The Disowned* (1829), which sees it as 'the heart of Avarice systematised'. The young Disraeli satirized middle-class pretensions in *Popanilla* (1827), describing how 'great moves to the westward were perpetual' in the capital, and how 'sumptuous squares and streets were immediately run up' in the West End to receive the burgeoning *nouveaux riches*. An excellent witty description of a new 'villa', Zero Lodge, will be found in M. J. Higgins's 'Jacob Omnium, the Merchant Prince' (1845). The story is worth reading for its good-humoured *exposé* of social shams ('Jacob Omnium, M.P., The Merchant Prince', *New Monthly Magazine*, Aug. 1845, 567–78).

[8] 'every class or combination of men have a strong propensity to get up a system of morality for themselves, that is conformable to their own interests . . .' (James Mill: 'The Formation of Opinions', *Westminster Review* (1826), vi. 255).

belonged the men of letters, who, free from patronage, began in the nineteenth century to be seen as members of a profession, from which it was possible to earn a legitimate and comfortable living. Shelley's 'unacknowledged legislators of the world' were, in fact, coming into their own.

Although the professional author was genuinely concerned for the lot of the poor and oppressed, he was sufficiently a man of his time to believe that it was legitimate to enjoy the fruits of his labours. Dickens, for instance, was an excellent man of business, investing throughout his career in steady Government, Russian, and Indian stock, railway paper and property. His bank accounts in his later years show such huge deposits that one could be forgiven for viewing him as a 'veteran Mammonite', and more than one American newspaper waxed satirical over his seeming financial acquisitiveness.[9] When he died in 1870 he left £93,000.[10] Anthony Trollope, towards the end of his autobiography, provides the reader with a list detailing the various sums he received from his writing, neatly and accurately totalling £68,939.17*s*.5*d*. This sum he regarded as 'comfortable, but not splendid'.[11]

One is not here trying to impute a kind of hypocrisy to these novelists in detailing their financial successes. As sons of their age, they are openly proud of their success and, more important, their readers, too, would have shared in their pride. Money of itself was free of moral reprobation: it was the misuse of money, the acquisition of wealth for its own sake and with no care for others, that constituted what these authors loved to call Mammon, or Baal, or the Golden Calf. Mammon afflicted the individual soul, so that it became blind to the sufferings of the poor; to renounce Mammon was to renew the humane spirit of man, and to redeem the afflicted individual. It was, then, a moral, rather than an economic or political limitation, that the foes of Mammon condemned.

[9] The substance of these remarks on Dickens is derived from M. V. Stokes, 'Charles Dickens: a Customer of Coutts & Co.', the *Dickensian*, Jan. 1972, 17–30.

[10] J. Forster, *Life of Charles Dickens* (7th edn., 1872), iii, p. 518. The sum, in present-day values, would exceed £1 million.

[11] A. Trollope, *An Autobiography* (1883); World's Classics edn. (1968), pp. 312–14. For an outstanding study of a Victorian author's awareness of his position in the expanding literary market of the time, see Dr James M. Brown's *Dickens: Novelist in the Market Place* (1982). It was reviewed by the present writer in the *Dickensian* (Autumn 1982), 169 f.

II

The nineteenth-century novelist who wished to reflect in his fiction the doings of the City and its luminaries found himself living in two worlds, physically and spiritually kept apart by Temple Bar, that barrier between the aristocratic West End and the bustling commercialism of the complex of institutions and businesses known loosely as 'the City'.[12] As an intellectual he would be at home in the scholar's study with Adam Smith, Ricardo, James Mill, and the other economic philosophers, and would usually react to them with varying degrees of hostility. But as a transmuter of real life into art he would have to forsake the study, and plunge into the thronging world of Threadneedle Street, Capel Court, and Lombard Street, where he would see at first hand the creators of wealth going about their business. If we are to understand the novelists' stock of conceptions and ideas relating to economic theory and capitalist practice in that age, it would be as well to remind ourselves briefly what those theories and practices were.

Examination of nineteenth-century economic theories reveals a hierarchy of ideas reaching back to the work of the Dutch jurist Grotius (Hugo van Groot), whose influential treatise *De Jure Belli ac Pacis* appeared in 1625. This work contained a chapter on Contract, with discussion of the causes determining the prices of commodities, and these early theories of value were adapted and developed a century later by Francis Hutcheson (1694–1746), who was appointed Professor of Moral Philosophy at Glasgow in 1729. Hutcheson's pupil was Adam Smith (1723–90), who himself occupied this chair in 1752.

In 1759 Smith published his *Theory of Moral Sentiments*. A noble and immensely readable work, it yet contains a framework of arguments to which the modern reader with any knowledge of the realities of nineteenth-century social life reacts nervously. Smith argued that as individual self-interest was permitted by the Creator, it would always be controlled in such a way as to bring about universal good. An enlightened society will thus leave man free to do what he feels impelled to perform, and by his so doing society as well as the individual will benefit. This is *laissez-faire* in the making, a credo which, in conjunction

[12] Wren's Temple Bar of 1670–2 was removed in 1878–9, but the expressions 'west and east of Temple Bar' continued to epitomize the old, aristocratic and the new, middle-class polarities of society.

with the refinements of Jeremy Bentham's 'greatest happiness' principle, would give rise to so many of the social injustices of the succeeding century.

Smith's economic thinking was strongly influenced by the work of the continental 'physiocrats', whose theories derived from those of the French physician François Quesnay (1694–1744). It was Quesnay and his followers who developed the concept of 'distribution of wealth', with its necessary creation of the idea of three 'classes': labourers, employers of capital, and landlords. Smith accepted this categorization and discussed it in his great work *Inquiry into the Nature and Causes of the Wealth of Nations* (1776). Here will be found essays on the product of labour and its distribution, the accumulation and use of indigenous goods, the progress of wealth in the principal nations, national revenue, and systems of political economy. And because Smith accepts Quesnay's concept of classes, we find in his work a habit of viewing the economic world as being exclusively dominated by contracts between parties bargaining from a basis of abstract equality. This habit he bequeathed to his successors, who were all men of the study, possessing in varying degrees the happy faculty of separating in their minds the realities of the master–servant relationship from the a priori propositions of an economic philosophy.

It was David Ricardo (1772–1823) who, in his *Principles of Economy and Taxation* (1817), postulated a fixed relationship between labour, capital, and landowner in his theory of economic rent. 'Rent' in Ricardo's sense was the share of production that accrued to the owner of land, as distinct from wages and interest, which fell to the labourer and the capitalist respectively. In the nineteenth century the equation of land and capital became so close that 'rent' tended more and more to mean interest on money invested. 'Rent', said Ricardo, was fixed by an immutable law. There is a class of land on the margin of cultivation which can produce only sufficient to pay for the labour and capital put into it. If any rent is charged, such land will fall out of cultivation. This idea serves as the base-line for rent, which rises or falls with the price obtained for the produce. A rise in prices will make it possible to cultivate land that has been lying idle; a fall will have the opposite effect. Because of this 'law' of rent, it followed, said Ricardo, that self-interest would force the labourer, the business man, and the landowner to harmonize their relationship, thus approaching the ideal economic unity.

Ricardo's rigidly mathematical theories created what may be termed 'economic man', an intellectual abstract conforming to patterns of behaviour not to be found in real society. Nevertheless, they brought about a philosophical harmony agreeable to abstract theoreticians, and also to those leaders of the new industrial age who may have desired a theory to justify their practices. Certainly the theory of economic rent dehumanized society in the minds of those who advocated it: men became mere agents of a single economic urge, that of competition; emotions and spiritual yearnings had no place in the constitution of 'economic man'.[13]

Smith and Ricardo have been chosen for special mention because they illustrate so effectively the characteristic limitations of their kind. Their economic theories are rooted in philosophy, and were developed and refined largely by other philosophers and by logicians. They assembled impressive compendia of facts, but used their data to erect basically a priori arguments. The philosophical mind is always concerned to achieve a totality or synthesis of the various branches of human experience and activity. Such an aim can remove its pursuers from the realities of life as it is being really lived around them. In the case of the economic philosophers, it was to lead to accusations of heartlessness, and even of malign opposition to individual human fulfilment.

Thus it was that the successors of Adam Smith—Thomas Malthus, with his theories of population control, Ricardo, and James Mill, who neatly united their ideas in yet another refined demonstration of mathematical logic—came to be seen by their opponents as a co-ordinated school. It would have been unjust to accuse these men individually of inhumanity, but it was possible to do so if one saw them banded together as 'Benthamites' or 'Utilitarians'. Jeremy Bentham had truly desired to achieve 'the greatest happiness of the greatest number' through extensive legal, political, and social reforms, and the same fundamental disinterestedness could be found in those Utilitarian Whigs, such as Lord Jeffrey and the Revd Sydney Smith, who propagated their views in the pages of Bentham's *Westminster*

[13] 'The theory is based on the assumption of competition . . . the landlord endeavours to obtain the highest rent he can, and the tenant the lowest . . . both are independent and intelligent agents . . . the landlord will not be influenced by kindly feelings, or political obligation, or long connection . . . the tenant produces with a single view to the sale of his produce, and . . . is able and willing to move, taking with him his improvements or their value, to any soil, or place, or trade, where he will be more favourably situated . . .'. L. L. Price, *A Short History of Political Economy in England* (3rd edn., 1900), p. 80.

Review.[14] It cannot, however, be denied that the general addiction of the philosophical radicals and their various sympathizers to theoretical solutions based on abstract concepts of egalitarian democracy led at times to an apparent hardness of heart in the face of individual suffering.

For those members of the newly-emancipated bourgeoisie who ever gave a thought to philosophies, the ideas of Utilitarianism no doubt appeared reasonable and just, as they in no way impinged upon the right to pursue affluence, a basic tenet of that 'enlightened self-interest' which constituted the entrepreneurial ethic.[15] However, among those who refused to see human beings in terms of statistics it was inevitable that a reaction to this consciously-held[16] rationalistic outlook should have occurred. Literary men were struck forcibly by the patent evils and inequalities surrounding them, and by the freedom afforded to those who pursued Mammon by the apparent justifications offered by Benthamism.

One must, however, exercise caution over literary responses to Utilitarianism, and not condition oneself to see the Victorian novelist as a revolutionary in the making. True sympathy for the underprivileged, and continuing protest at their inhuman treatment, was combined with an innate conservatism, which looked, not to state intervention, but to private philanthropy for the alleviation of society's ills. Reform was to come from the heart, from individual compassion, rather than from the rational application of soulless 'laws' by government or bureaucracy.

It was usually the upper classes, suitably changed in heart, who were seen by these writers as the legitimate initiators of reform. There is an unwillingness to envisage the working man taking part in the creation of his own destiny. In *Hard Times* (1854) Dickens depicts the

[14] Jeffrey and Smith, of course, are more celebrated for having founded, with Brougham, the *Edinburgh Review* in 1802. *The Westminster Review* was established by Bentham and James Mill in 1824 as the avowed organ of philosophical radicalism.

[15] 'The economist was pleased to see the conclusions of his science confirmed by the approval of the men of practical life, while the doctrine which made individual selfishness the sole governing factor in the progress of humanity was equally seized upon by the middle classes as the best means of advancement of their class interests.' (H. Jansonius, '*Some Aspects of Business Life in Early Victorian Fiction*', doctoral thesis, Amsterdam, 1926, pp. 13–14.)

[16] W. R. Greg openly acknowledged the difference between Benthamite philanthropy and the personal philanthropy of the type exercised by, for instance, Angela Burdett-Coutts: 'There are two classes of philanthropists, the feelers and the thinkers—the impulsive and the systematic.' (Review of *Alton Locke,* the *Edinburgh Review* (1851), xciii, 3–4.) Greg was a 'thinker'.

trade union as a body ultimately unable to help the operative, and shows the union organizer as an unscrupulous agitator. Kingsley, the Christian Socialist, helped in the founding of co-operatives, but advised the Chartists in April 1848 to leave reform to the upper classes.[17]

Although one will find few if any apostrophes to Mammon in Dickens's works, one does see a continuing reaction to the social problems of the time, and an awareness of the determining forces underlying them. Dickens is always conscious of the debilitating effects of 'money-getting', and of the fact that money can play a divisive role, isolating man from man, and generating inhumanity, selfishness, and hypocrisy. Bound up in his mind with the misuse of money was the wrong-headed application of Utilitarian ideas to justify harshness and indifference towards the humble and weak.

Perhaps his most celebrated satire on 'Benthamism' is *The Chimes*, the Christmas Book for 1844. The statistical approach to social problems is mercilessly guyed in the person of Mr Filer, whose name betrays his enslavement to political economy. He it is who expresses publicly his indignation at Trotty Veck's humble bowl of tripe. 'I find that the waste on that amount of tripe, if boiled, would victual a garrison of five hundred men for five months of thirty-one days each, and a February over. The Waste, the Waste!'[18] Further statistics 'prove' that the poor ticket-porter is a robber, who snatches tripe 'out of the mouths of widows and orphans'.

In one sense, Dickens is not exaggerating the Utilitarian enchantment with figures. The Benthamite *Westminster Review* in June 1844 had found fault with Bob Cratchit's good fortune in having turkey and punch at Christmas: 'Who went without turkey and punch in order that Bob Cratchit might get them — for, unless there were turkeys and

[17] In the world of *belles-lettres* Thomas Carlyle must be singled out for his unrelenting attacks upon Mammon in general and Benthamism in particular. As early as 1829 he was complaining that 'Men are grown mechanical in head and in heart as well as in hand' (*Edinburgh Review* xlix, June 1829, p. 440 f.), and his strictures are best exemplified in the rhetorical exuberances of *Past and Present* (1843). Society had become fragmented into self-seeking interest groups, 'Midas-eared Mammonisms', and the poor were left to scramble for subsistence according to 'the law of the Stronger'. Carlyle's life-long broodings on this Mammon theme are well illustrated in Caroline Fox's *Memories of Old Friends; 1835 to 1871*, where, in correspondence and personal meetings, Carlyle is shown increasing in pessimism and despair over the way society was moving into a fragmented state of class antagonisms. See H. N. Pym's edition of the *Memories* (1883), especially pp. 219, 223, 413, 414.

[18] C. Dickens, *Christmas Books* (Penguin English Library), ed. Michael Slater, i, p. 166.

punch in surplus, someone must go without—is a disagreeable reflection kept wholly out of sight . . .'. Mr Filer and his homily on tripe are said to be a direct rejoinder to this perverse and amazingly mean-spirited critique of *A Christmas Carol*,[19] and one is not surprised to find that *The Chimes* progresses into a developed attack on most Benthamite economic assumptions.

The influence of the economic philosophers on creative minds was profound, and various forms of acceptance and rejection have been fully discussed from the literary point of view in Louis Cazamian's classic *Le Roman Social en Angleterre* (1903). Very few people today who are interested in the social aspects of nineteenth-century literature will be unacquainted with at least the names of these early economists, if only because they are linked so closely with literary condemnations of self-interest, oppression of the weak, and the corrupting power of wealth. And it is here that we need to pause and ask ourselves whether a true picture of the institutions of capitalism can be obtained either from the economists themselves or from their literary detractors.

There was, in fact, another class of economic writers who may be termed the commercial and fiscal historians. These men, often fully experienced in the realities of industry and commerce, performed an invaluable service in recording in detail the business history of their age. Though at times limited by prevailing economic belief in 'laws' and 'patterns', they were able, unlike the majority of philosopher-economists, to confine their descriptions to single financial or commercial events, often providing sound, and more often than not completely correct, analyses of the causes of specific phenomena. It is to these men that one looks for the practical view of what occurred on the wrong side of Temple Bar.

Among these commercial historians David Morier Evans deserves particular mention. He was a financial journalist, and his three meticulous works, *The Commercial Crisis 1847–1848* (1849), *Facts, Failures and Frauds* (1859), and *The History of the Commercial Crisis 1857–1858* (1859), provide an indispensable mine of facts covering a very important decade of commercial life in England.[20] Similarly,

[19] Slater, p. 262, n. 7. Forster admired *The Chimes* immensely, but Bulwer, also a friend of Dickens, thought that its moral was 'untrue and dangerous', and that its 'fierce tone of menace to the rich is unreasonable and ignorant'. See Slater, p. 139, including footnote 4.

[20] David Morier Evans (1819–74) was a professional financial journalist. For many years assistant City correspondent of *The Times*, he later wrote the money articles in

no close discussion of nineteenth-century commerce could avoid reference to the Russia-merchant Thomas Tooke's monumental *History of Prices*, which was issued in instalments between 1838 and 1857, or to the works of the merchant-historian H. R. Fox Bourne, *English Merchants* and *The Romance of Trade*, which appeared in the sixties.

It will be remembered how Dickens's Mr Gradgrind in *Hard Times* (1854) was enthralled by 'facts': 'Now what I want is, Facts . . . Facts alone are wanted in life. Plant nothing else, and root out everything else.' (Ch. 1.) These were the 'dehumanizing' facts adduced to support the theories of Ricardo and the Utilitarians. Of a rather different nature was the detailed factual information appearing in the tables and appendices of the works of the commercial historians. There was a definite movement from the thirties onward away from abstract theorizing and towards statistical enquiry. The statistical department of the Board of Trade was established in 1832, and the Registrar-General's Department in 1838. The 'facts' collected and published by these bodies, and by the expanding insurance companies of the period, were of the type that revealed in the various government reports of the era the true, rather than the theoretical, state of the nation. As G. M. Young tells us, the influence of the Blue Books of the Royal Commissions after 1832 provided 'fresh topics for novelists and fresh themes for poets'.[21] It is with the idea of teaching practical lessons and revealing mistaken practices and abuses that the commercial historians assemble their statistics, and in this they are at one with the earnest social reformers of their time.

Where, then, should the modern reader turn to find a valid depiction of nineteenth century capitalism? The professional historian views with great caution the factual information about social or political conditions

the *Morning Herald* and the *Standard*, and was a frequent contributor to the *Bankers' Magazine*. In his writings he carefully avoids theories, and confines himself to detailed chronicling of events, from which he draws *ad hoc* conclusions. See in particular his statement of method in *The Commercial Crisis 1847–1848* (1849), p. ix.

[21] G. M. Young, 'Portrait of an Age', *Early Victorian England* (1934) ii, p. 441. Ricardo, Bentham, Malthus, and James Mill were all dead by 1836. The new generation of economists tended to be more aware than their predecessors had been of the value of statistical bases for enquiry. Mention may be made of John Ramsey McCulloch, whose *Dictionary of Commerce* appeared in 1845, and of Leone Levi, author of *A History of British Commerce* (1872). McCulloch (1789–1864) was Professor of Political Economy in the University of London (1828–32), and Comptroller of the Stationery Office (1828–64). His classic *Statistical Account of the British Empire* appeared in 1837. Levi was Professor of the Principles and Practice of Commerce at King's College, London.

provided in the novels of the period, because he is aware of such factors as literary prejudice, an author's own conception of his standing in the hierarchies of his own epoch, and the selectivity of his creative responses to people and situations. W. O. Aydelotte aptly summarized the historian's approach when he remarked that 'the attempt to tell the social history of a period by quotations from its novels is a kind of dilettantism which the historian would do well to avoid'.[22]

With that salutary warning in mind, I have attempted in this book to examine fictional creations in the light of the extremely well documented commercial and economic history of the nineteenth century. To do so helps towards a realization that the novelist as a creative artist was not wholly motivated by reactions to the philosophical theses of Bentham, Malthus, James Mill, and the rest. The novelist was not primarily an academic, engaged in dispute with Scottish dons, but an inventive craftsman, whose particular genius would utilize the preoccupations and institutions of his time for his own inner purposes.

III

Disraeli, in *Coningsby* (1844), accurately characterized the new spirit of commercial and industrial enterprise prevailing after 1815 as:

> . . . the most extensive foreign commerce that was ever conducted by a single nation . . . an internal trade supported by swarming millions, whom manufactures and inclosure-bills summoned into existence; above all, the supreme control obtained by men over mechanical power—these are some of the causes of that rapid advance of material civilization in England, to which the annals of the world can afford no parallel. (II, Ch. 1.)

Hand in hand with this industrial and commercial expansion went a rapid advance in the recognition of credit as the prime mover of commerce. Daniel Webster asserted that credit had done more to enrich nations than all the mines in the world, and Fox Bourne agreed:

> The legal currency, whether gold, silver or bank notes, is only a sort of pocket-money in comparison with the real currency of trade. It serves for the smaller sort of retail purchases . . . but the great merchant has not in his possession all through his lifetime actual money equal in amount to the paper equivalent of money that passes through his hands every day of the week.[23]

[22] W. O. Aydelotte: 'The England of Marx and Mill as Reflected in Fiction', *Journal of Economic History*, 88 (1948), 43.
[23] H. R. Fox Bourne, *English Merchants* (1866), 2nd edn. 1886, p. 474.

By 'credit' we are to understand that complex of paper transactions which does the work of money without the actual passing of currency. The system probably had its origins in deferred payments for retail transactions, but the greatest impetus to its acceptance was the rapid growth of foreign trade in an age of slow transport and communication. For example, a merchant in Madras could not expect quick payment for the goods he dispatched to London. Consequently he would draw up a 'bill of exchange', ordering the recipient to pay for the goods three, four, or six months 'after sight' of the bill. Thus the foreign bill of exchange, usually ninety days from sight, came into being, soon to be followed by the inland bill of exchange for home transactions. These bills quickly became a form of currency, endorsed, and used to pay debts to third parties; or they could be taken to a bank for discounting. The bank would pay the current holder the value of the bill, less the amount of interest it would lose until the bill became due. Specialist firms, known as discount-houses, developed to deal exclusively with this form of paper transaction.

The cheque, essentially another form of credit, developed from the bill of exchange, and by the beginning of our period it was most unusual, if not eccentric, for any business transaction to involve the physical transfer of gold and silver. Mr Dombey, the Cheeryble brothers, Ralph Nickleby, Ebenezer Scrooge, and all other true men of business would have been entirely familiar with the many banks, discount and accepting houses, and the whole machinery of credit that constituted 'Lombard Street', or the 'Money Market', as the system was often called.

Central to the City, and virtually side by side in the triangle formed by Throgmorton Street, Threadneedle Street, and Old Broad Street, lay the Bank of England and the Stock Exchange, the prime facilities for the accumulation of savings and capital. The Bank had been established in 1694 at the suggestion of the Scottish financier William Paterson,[24] as a means of collecting and managing public loans for the reduction of the National Debt. A fund of £1,200,000 was raised in a period of ten days, and the Bank duly received its Royal Charter as a joint-stock company, with permission to deal in bills of exchange, bullion, and forfeited bonds.

[24] Born in Dumfriesshire in 1658, Paterson left Scotland in 1685, and lived for five years in the Bahamas. He returned to England in 1690, and engaged in business, founding the Bank of England in 1694. He was the author of the ill-fated Darien Scheme. He returned from Darien with the few survivors in 1699, and died in 1719.

In 1709 the directors of the Bank secured the passage of an Act of Parliament that forbade the establishing in England of any joint-stock note-issuing banks in which there were more than six proprietors. The Bank thus became the only effectual joint-stock bank in the kingdom, a monopoly which it jealously guarded until 1826, when joint-stock banks outside the capital were authorized.

The early decades of the nineteenth century were precarious ones for those who earned their living in the City; and the Bank, like any other commercial institution, had to accommodate itself to panics and crises requiring drastic and often dramatic solutions. Fundamentally, though, no financial house was safer than the Bank of England. Although it was never in our period a mere branch of the Treasury, a mutually-accepted if tacit interdependence had developed between the two institutions, and the Bank rightly epitomized stability and confidence. Lodged in its imposing Roman–Corinthian edifice built by Sir John Soan in 1800, it was a dominating and familiar sight to the denizens of the City throughout the period, a vast complex of over two hundred offices, covering an area of 124,000 square feet. It was staffed by over a thousand clerks.[25]

Fox Bourne allows us to see the Victorian attitude to the Bank in the following description:

> The governor, deputy-governor, and twenty-four directors . . . regulate, to a very great extent, all the commercial affairs of England, and even of every other country . . . their decisions have vastly more influence upon the happiness and activity of men than any resolutions of cabinet councils, or any proclamations of kings or emperors.[26]

Those concluding words, exalting the Bank of England above the legislatures of the nations, are a fairly typical response from both Victorian chroniclers of finance and those novelists who mention it in their works. Its very real power is acknowledged with a pride not unmixed with awe. However, the Bank's function in literature could never be more than that of occasioner of a moral aside, or a light squib directed against its personnel. The novelist would not find naughtily exciting, stimulating, or sensational material in the daily workings of such an open institution as the Bank of England.

[25] These clerks were the highest type of City executive in the early decades of the nineteenth century. Clerks in other banks and financial institutions were the servants of their masters. (See R. H. Mottram: 'Town Life', in *Early Victorian England*, i, p. 179.)

[26] H. R. Fox Bourne, *The Romance of Trade* (n.d.), p. 83.

More fruitful material for literary invention lay with the Bank's near neighbour the Stock Exchange, and with the great number of private financial houses of various types to be found in the purlieus of the Bank, throughout the City, and scattered through the cities and county towns of England. Succeeding chapters will show how nineteenth-century novelists regarded these institutions, and what literary use they made of them. They will also reveal the extent of the novelists' actual knowledge of the workings of financial institutions, and the type of responses they made, both to literary traditions about the City and to the commercial events of their own time.

During this age of commercial and industrial expansion, investment in the growing institutions of capitalism was a national pastime, affecting all classes of society. As the concept of credit became generally understood, people began to look beyond their individual stocks of guineas and sovereigns for opportunities to achieve paper affluence. The prudent would be content with the 'Funds', secure government paper,[27] knowing that the modest interest would always be paid, whatever the financial or political climate. The naïve, the ignorant, the deceived, and the greedy would find in the nineteenth century many opportunities to 'speculate', or gamble, in enterprises where fortunes were either fallacies or fictions of accountancy.

The nineteenth century was a commercially literate age. The workings of the Stock Exchange, the means of obtaining discount on a bill, such things were widely known, and novelists who made use of paper transactions in their works knew well that all but a few of their readers would understand to what they were referring. The 'game of speculation' was played by all classes, and to make one's fortune was considered a laudable aim. During the two principal fevers of speculation connected with the Company and Railway Manias, thousands of quite humble people ventured their modest savings or mortgaged their incomes in investments which they hoped would make their fortunes, and those too poor to do so wished that they could. Manufacturers, merchants, factors, bankers, people on fixed incomes, retired half-pay officers, governesses, widows, trustees of orphans'

[27] The term 'paper' is used loosely in commerce for such items as government bonds, share certificates, bills of exchange, etc., the immediate context indicating the type of 'paper' to which reference is being made. It is so used throughout the present work.

funds, shopkeepers, aristocrats, and gentry, all rushed in those years to the stockbrokers to claim their stake in the new Age of Gold.[28]

This broad cross-section of society constituting the 'investing public' furnished the novelist with almost unlimited material for dramatic representation and moral comment. Miss Mulock, for instance, describing the expected fall of Jessop's Bank in *John Halifax, Gentleman*, is careful to insist on the representation of all ranks in the crowd before the closed doors of the bank: 'It included all classes, from the stout farmer's wife or market-woman, to the pale, frightened lady of "limited income" . . . from the aproned mechanic to the gentleman who sat in his carriage at the street corner . . .' (Ch. 31).

Novelists were quick to dramatize the quite genuine plight of the widow and orphan, and to castigate the heartless instigators of speculation. Nicholas Nickleby's father is described as having been one of 'several hundred nobodies' ruined by speculation, and Montague Tigg, the fraudulent insurance promoter in *Martin Chuzzlewit*, sneers at his dupes as 'a little tradesman, clerk, parson, artist, author, any common thing you like' (Ch. 27).

The unfortunate investors themselves did not escape the moral strictures of the novelists, and are frequently taken to task for yielding to the blandishments of the 'golden promises' held out to them at times of financial buoyancy. Such reactions seem to arise from a sense of impotence, a feeling that speculator, investor, principal, and indeed novelist, were ultimately powerless to resist the force of each commercial upheaval as it afflicted society. For instance, in Thackeray's *The Great Hoggarty Diamond* (1841), there is an examination in bankruptcy following the flight of a swindling company promoter. The words of the commissioner are directed as much against the deceived investors as the delinquent business man:

> 'If you had not been so eager after gain, I think you would not have allowed yourself to be deceived . . . Directly people expect to make a large interest, their judgement seems to desert them' . . . 'But what's the use of talking?' says Mr. Commissioner in a passion: 'here is one rogue detected, and a

[28] See for instance the *Annual Register* for 1824 (c. 3), which effectively epitomizes those who scrambled for new ventures in the mania of that year:

> All the gambling propensities of human nature were brought into action . . . princes, nobles, politicians, placemen, patriots, lawyers, physicians, divines, philosophers, poets—intermingled with women of all ranks and degrees, spinsters, wives, and widows, to venture some portion of their property in schemes of which scarcely anything was known except the name.

thousand dupes made; and if another swindler starts tomorrow, there will be a thousand more of his victims round this table a year hence; and so, I suppose, to the end.' (Ch. 12.)

It will become depressingly clear in later chapters of this book that Thackeray's pessimism about players in the game of speculation was to be fully justified.

IV

Two characteristics of the nineteenth-century financial scene encouraged violent bursts of imprudent speculation and widespread fraud and mismanagement. One was the imperfect state of company law, particularly before 1844, and the other was the cyclic pattern of commercial crises occurring every ten years or so, virtually predictable, and seemingly unavoidable.

Since the development of the mercantilist system in England there had been two kinds of company, the regulated and the joint-stock. Regulated companies were few in number and large in size, enjoying with government blessing a monopoly of trade with distant countries. The Russia, Levant, and Africa companies were examples of such regulated partnerships.[29]

Joint-stock companies were associations in which the capital stock of the members—called at first 'proprietors' and later 'shareholders'—was amalgamated. Each shareholder shared in the profits made with this 'joint stock', in proportion to the amount contributed, and was liable only for the amount he had put into the undertaking.

Joint-stock companies could be legally established in two ways. Incorporation by Royal Charter conferred upon them the highest social recognition for probity—the East India Company had been thus incorporated in 1600, and the Bank of England in 1694. The alternative was incorporation by Act of Parliament, the method sought by the numerous railway companies founded in the early Victorian period.[30]

[29] A full account of the then prevailing English company constitution will be found in J. R. McCulloch, *A Dictionary of Commerce* (1845) under 'Companies'.

[30] Incorporation by Act of Parliament was statutory for companies which, like those for promoting railways, sought to establish a monopolistic privilege not claimed by the Crown.

Incorporation by either method was well beyond the means of ordinary trading partnerships,[31] so that the majority of joint-stock companies remained unincorporated, and thus unrecognized in the Statute Law. Whatever their size or prestige, such companies were merely notional partnerships. An Act of 1825 made joint-stock companies amenable to the Common Law, but at best this enabled an individual shareholder to sue a defaulting client, or himself be sued for default. It was still not possible to hold the 'company' responsible for its conduct as a single entity.[32]

A belated attempt to recognize the realities of business life came with an Act of 1837 (1 Vic., c. 73), which declared that it would be 'inexpedient' to incorporate joint-stock companies by Royal Charter, but 'expedient' to grant them, on request, Letters Patent to sue or be sued in the name of one or two officers of the company. Companies electing to be registered could do so at government enrolment offices.

By leaving the option of registration to the promoter, the Government allowed him the choice either of working openly under rudimentary control, or of working as secretly and as irresponsibly as he wished. The fraudulent promoter, by omitting, quite legally, to register his company under the Act, could ensure absolute secrecy for his schemes, rendering himself amenable only to common law action under the Act of 1825. As his sole object was to fool as many dupes as possible before absconding with his loot, he was undaunted by the prospect of civil actions for restitution.

It is as well to remember that shareholders of swindling joint-stock companies could also be victims of a fraud perpetrated by their own board, and that such frauds were common throughout the century. Fraudulent promoters and directors shrewdly calculated that victims would not risk further money in resorting to the law, even after Acts of 1844 and 1855 had brought full statutory recognition to joint-stock companies.[33] This imperfect and slow development of company law

[31] It cost £70,000 to obtain Parliamentary permission to build the Liverpool to Manchester line. The thirty-five railway Bills passed in 1836 involved an estimated expenditure of £17,500,000. Dickens speaks of 'a Private Bill, which of all kinds and classes of bills is without exception the most unreasonable in its charges' (*Martin Chuzzlewit* Ch. 27).

[32] See for example the case (*Buck* v. *Buck*) in Wordsworth, *Law of Joint Stock Companies* (1842), p. 21.

[33] For frauds perpetrated against shareholders see J. E. T. Rogers, 'The Joint-Stock Principle in Capital', in his *Industrial and Commercial History of England* (1892), p. 145. The Companies Act of 1844 made registration compulsory, together with annual

provided in reality many dramatic models for the novelists' tales of chicanery and fraud in all branches of commerce, as succeeding chapters will amply demonstrate.

The cyclic pattern of ten-year commercial crises exercised the minds of all who were engaged in business, whether as merchant, banker, speculator, or investor. It was observed that from the beginning of the century periods of prosperity, with money cheap in the market, were followed by general scarcity and a financial crisis in the City.[34] These events seemed to occur in cycles of roughly ten years. Businessmen, economists, and the public came to accept this pattern as yet another 'law' of economic life, and while theories were from time to time postulated, no scientific explanation of the cycle was ever reached.[35] Precariousness was thus part of the system, and bankers and merchants were conditioned to accept at least the possibility of enormous losses consequent on depression and diminishing credit as the crisis of each cycle approached.

The prevalent attitude of City figures towards the cycle is well typified by Fox Bourne's observations in *The Romance of Trade*. Some commercial instinct impels him to regard the theory as 'fanciful', but he does not categorically deny the possibility of its being true:

> During the forty years following upon 1825, English trade throve wonderfully; but there has been a monotony of variety in its progress, tempting fanciful observers to propound a law of financial tides and storms. After a crisis there is a lull of a year or two . . . Then speculating energy revives and steadily gains force during seven or eight years, until it develops into a mania, lasting for about a year, and ending in another panic. (Ch. 11.)

publication of a balance sheet, thus affording considerable safeguards to creditors. Limited liability for shareholders was made possible by the Companies Act of 1855. These two Acts finally brought full statutory recognition for joint-stock enterprises.

[34] 'The City had fallen into a habit of crises so nearly every ten years that there was some idea that such events were inevitable. Boom and depression certainly alternated . . . with a regularity that gave grounds for that impression' (R. H. Mottram, 'Town Life', p. 182).

[35] Both classes of economist failed to observe the cycle scientifically. Tooke, in his *History of Prices*, assembled every fact available, but seemed to have had no general picture in his mind of the trend of prices. (See Tooke, ed. T. E. Gregory (1928), p. 21.) J. S. Mill's *Political Economy* (1848) was similarly defective. Stanley Jevons was the first professional economist to attempt a scientific explanation of the wave movement, postulating a correlation between trade fluctuations and harvest fluctuations. The latter, he said, could be affected by the sun-spot cycle, which lasts eleven years. William Smart, an advocate of Jevons's theory, states his own shrewdly argued views on the cycle in his *Economic Annals of the Nineteenth Century* (1910), I, 605 f.

When the Victorian novelist decided to use a business theme in his fiction, he brought to that theme a complex of attitudes and ideas conditioned by the material examined so far. He would have some awareness of prevailing economic theories, their use and misuse; he would sense the growing power of capital and the rapid burgeoning of commerce. He would be only too well aware of the precariousness of investment in an age of imperfect company regulation, and of the dramatic nature of the many uncertainties produced by the various cycles of boom and crash. And, in an age given to moralizing, he would find ample material in the regular public frenzies of mindless, mesmeric speculation and abandonment to paper dreams.

All these things furnished the novelist with fertile material for dramatic or moralistic fiction. But the creative artist brought to bear upon his work an awareness that was unique to his *métier*. This was a literary tradition of suspicion about the doings of 'money-getters' which had its genesis at least as early as the seventeenth century. Mercantile pursuits were in the main kindly regarded, as being essential to the nation's well-being, honourable in performance, and venturesome in spirit. But those who dealt solely in money, or who manipulated commercial undertakings purely out of 'greed for gain', were subject to varying degrees of literary invective. It was, perhaps, inevitable that the defalcations of the few should condemn the many, producing in literary works violent distortions of sober, if not very dramatic, truth. In seeking to gain some insights into nineteenth-century business life from the novels of the period, their readers need to tread a wary road.

1

Brokers and the Stock Exchange

The second great City institution after the Bank of England was the Stock Exchange. Then as now a private association, it was the meeting-place of the numerous stockbrokers and stockjobbers, who, by the 1840s, had achieved an essential and 'respectable' standing in the City. Literary depictions of these men, however, are frequently coloured by an antagonistic tradition stretching back to the seventeenth century, and it is important for a full understanding of such depictions to examine the genesis and progress of that tradition.

It was King William III who virtually established what came to be known as the 'National Debt' by introducing a new system of interest-bearing paper known popularly as 'the Stocks' or 'the Funds'.[1] The settled political climate after 1688 caused many people to abandon the practice of hoarding their money in chests, or of paying goldsmiths and merchants to guard it for them, and to turn instead to buying and selling not only Government paper, but the private stock of the chartered trading companies. The trade of 'stock-jobbing' arose from the tendency of such public stocks to fluctuate in value.

The terms 'jobber' and 'broker' are frequently used in novels of the period synonymously, and usually with pejorative connotations. The terms are, however, indicative of distinct though closely allied professions. The jobber is a man of considerable substance who holds a wide range of stocks and shares which he is ready to sell; he is also able to buy other paper that is offered to him on satisfactory terms. The broker, on the other hand, is the agent of others, commissioned by them to buy and sell stocks and shares on their behalf.

The second principal activity of the jobbers and brokers of the Stock Exchange and its environs was to encourage, at times when money was plentiful, the promotion of new companies, in the paper of which they could deal at a profit. It was in this particular activity that

[1] 'Stock' was originally an abbreviation for capital stock, and was applied to the shares of the East India Company and the other seventeenth-century trading companies. 'Fund' signified the tax or fund set apart as security for the payment of principal and interest; later the term came to mean the loan itself, not the security.

the genesis of the antipathetic literary tradition lay. Encouragement of rash speculation led more often than not to a panic and wholesale losses, and this practice had a pedigree stretching back to the first appearance of jobbers in England. Macaulay, with personal memories of the speculative manias of 1825 and 1845 to colour his selection of material, vividly details the kind of enterprise encouraged and fomented by the early jobbers of the seventeenth century:

It was about the year 1688 that the word stockjobber was first heard in London. In the short space of four years a crowd of companies, every one of which held out to subscribers the hope of immense gains, sprang into existence; the Insurance Company, the Paper Company, the Lutestring Company, the Blythe Coal Company, the Swordblade Company.[2]

This sudden flurry of financial activity arose from increased financial prosperity and social stability, when redundant capital sought new outlets. Chicanery informed these activities from the beginning. Macaulay is again thinking of recent events when he writes:

It was much easier, and much more lucrative to put forth a lying prospectus announcing a new stock, to persuade ignorant people that the dividends could not fall short of twenty per cent, and to part with five thousand pounds of this imaginary wealth for ten thousand solid guineas, than to load a ship with a well chosen cargo for Virginia or the Levant. (III, Ch. 19.)

As early as 1692 Thomas Shadwell had written a comedy, *The Stock Jobbers*, which contains mock-serious discussions among canting Puritans about the respective merits of a 'Mousetrap Company' and a 'Flea-killing Company'. The play is the earliest, and one of the most effective, exposés, not only of dishonest jobbers, but of a new public hypocrisy that speculated with one hand and condemned speculation with the other.[3] The financial conditions described here, the public reactions, and the response of literary men, would be many times repeated in the two centuries that followed.

[2] T. B. Macaulay, *History of England*, iii (1855), Ch. 19. Macaulay states in the same chapter: 'A mania of which the symptoms were essentially the same with those of the mania of 1720, of the mania of 1825, of the mania of 1845, seized the public mind.' Born in 1800, he had personal recollection of the 1825 and 1845 manias, and it is partly because of this that he compiles a list of company names in the passage quoted. Bubble companies floated during 1825 frequently gave themselves ludicrously verbose names, which were burlesqued by novelists of the period.

[3] *The Volunteers: or the Stock Jobbers* was published posthumously in 1693.

Shadwell's play was a trenchant satiric reaction to the first 'company mania' of any magnitude to affect and excite English society. Most famous—or infamous—of such manias, in which the jobbers and brokers played a crucial part, was that known as the South Sea Bubble. The South Sea Company, promoted in 1711 by Robert Harley, Earl of Oxford, as a Tory counter-balance to the Whig-dominated Bank of England, was charged with the task of colonizing and trading with the western side of South America. The first trading-vessel did not leave London until 1717, but in the interim a busy trade was carried on in the Company's shares. The Company soon rivalled the Bank of England, and offered to advance the Government a loan of £2 million. During the resultant strife between the two institutions, there was a furious trade in shares, encouraged and in part fomented by the stockjobbers. Public and governmental confidence was invested in the South Sea Company, and in April 1720 an Act transferred the whole National Debt of £30 million to this colossal institution. This was the highpoint of a madness that would seize the country again in the first half of the nineteenth century.

The apparently incalculable paper wealth of this company had the effect of increasing the paper value of other chartered companies, so that, for instance, East India Company stock rose in 1720 from £100 to £445. It would appear that virtually everyone with money saved wished to invest, and in this year of the crash over two hundred 'bubble' companies were floated, with an aggregate paper capital of over £300 million.

Public confidence is always essential for the success of any commercial enterprise. When rumours of quarrels among the Company's directors reached the public, that confidence was lost, with the sudden and dramatic collapse of the Company, and of the hundreds of questionable undertakings that had grown around it. In August 1720 one South Sea share was worth £1,000; in September the shares could not be sold for £150. A great number of people were utterly ruined by this collapse, and for many years following, public confidence in the commercial institutions of the country was badly shaken. Many of the smaller 'bubble' companies were found to have been fraudulent; even worse, government ministers, including the Chancellor of the Exchequer, Aislabie, were seen to have used their official positions to profit from the gamble. Stanhope's faction fell from office, to be replaced by Walpole, and the new administration immediately resumed the National Debt, investing it once again in the Bank of England.[4]

Walpole characterized this mania in the following words:

> the great principle of the project was an evil of first-rate magnitude. It was to raise artificially the value of stock, by exciting and keeping up a general infatuation; and, by promising dividends out of funds which could never be adequate to the purpose, it would hold out a dangerous line to decoy the unwary to their ruin, by making them part with the earnings of their labour for a prospect of imaginary wealth.[5]

The rage for speculation that led to this enormous commercial convulsion was in great measure provoked by jobbers, 'hundreds of knaves to whom thousands of fools were willing dupes', as Fox Bourne puts it.[6] It is not surprising, therefore, that the role of the stockjobber, in conjunction with that of the directors of the Company as instigators of the project, was seen as being central to the mischief and its calamitous results. In the years following the fiasco, poets, essayists, and novelists kept the memory of both the principals and the jobbers alive in satire and invective. Particularly interesting are the reactions of Jonathan Swift, as they furnish pointers to the various ways in which writers in the nineteenth century were to respond to similar financial types and situations.[7]

Swift's satiric poem *The South Sea Project* appeared in 1721. His principal target is the board of directors, whom he damns as rogues and bloodsuckers, luring hapless investors with golden promises:

[4] See in particular L. Melville, *The South Sea Bubble* (1921), *passim*. Walpole was only too willing to prosecute the directors with exemplary violence. £2 million was raised from the confiscation of their property, and that of others peripherally connected with the Company. Edward Gibbon (1737–94), who was descended from one of the directors, complained that they had been unjustly, if not unconstitutionally, treated, 'condemned, absent and unheard, in arbitrary fines and forfeitures' (*Miscellaneous Works*, i (1796), p. 11).

[5] Quoted in Fox Bourne, *The Romance of Trade*, p. 314.

[6] Ibid., p. 317.

[7] There were a number of literary responses to the Bubble, primarily of an occasional character. Pope complained that

> . . . judges job, and bishops bite the town,
> And mighty dukes pack cards for half-a-crown;

in less exalted vein, street ballads and coffee-house epigrams satirized the Mania, the following being one example:

> The greatest ladies thither came,
> And plied in chariots daily,
> Or pawned their jewels for a sum
> To venture in the Alley.

(See Fox Bourne, *Romance of Trade*, pp. 314–5.)

Thus, by directors we are told,
'Pray, gentlemen, believe your eyes;
Our ocean's cover'd o'er with gold,
Look round, and see how thick it lies.'

Closely linked in Swift's mind with these 'devouring swine', as he calls them, are the jobbers, waiting like carrion for the floundering 'shipwrecks', so that they can 'strip the bodies of the dead'. The promises of these vultures will be equally illusory as those of the directors:

So cast it in the Southern seas,
Or view it through a jobber's bill;
Put on what spectacles you please,
Your guinea's but a guinea still.

As intermediaries, the jobbers will be able to make corrupt private fortunes by not appearing as principals in transactions:

When stock is high, they come between,
Making by second-hand their offers;
Then cunningly retire unseen,
With each a million in his coffers.

Swift is, of course, justified in his contempt of irresponsible directors and dishonest jobbers; at the same time, though, his satire, like that of Shadwell earlier, is an element of a literary tradition that would colour the attitudes and writings of men of letters in the succeeding century. While commodity merchants such as Dickens's Mr Dombey are usually granted personal and commercial integrity, those whose merchandise is money are seen as 'shady', or 'shaky', or lacking in some personally desirable quality.[8] Also, while merchants held a very respectable rank in society,[9] company promoters and jobbers were

[8] For instance, in *The Dog and Thief* (1726), Swift accuses stockjobbers of trying to bribe freemen for their votes:

The stockjobber thus from 'Change Alley goes down,
And tips you the freeman a wink;
Let me have but your vote to serve for the town,
And here is a guinea to drink. (ll. 9–12.)

[9] 'In this body of clerical activity . . . we have the central distinctive type of our period . . . Above it . . . were the City merchants, who made the Victorian age a world-wide instead of an insular phenomenon, Dombeys and Sedleys, Cheerybles and Clennams, East and West India merchants, exporters, shippers, dealers on commission, and correspondents of foreign houses.' (R. H. Mottram, 'Town Life', in *Early Victorian England* i, Oxford, 1934, pp. 180–1.)

early stigmatized as vulgar and pretentious. For example, in his *Pastoral Dialogue* of 1727, Swift placed these words in the 'mouth' of Marble Hill, the Countess of Suffolk's house at Twickenham:

> Some South-Sea broker from the city
> Will purchase me, the more's the pity;
> Lay all my fine plantations waste,
> To fit them to his vulgar taste . . .[10]

This image of 'vulgar taste' as characteristic of the commercial class grew stronger in the nineteenth century, as the power bases of the country began to polarize around the conflicting spheres of landed and commercial interest; and while men of Swift's rank would find it easy to stand aloof, the nineteenth-century author would often be unsure of his own standing, or uncertain to which class he truly belonged. However, the force of literary tradition, coupled with a reaction after about 1830 from the various inhumanities apparently sanctioned by Benthamism, made it easier for him to castigate 'moneyed men', not merely as swindlers, or parvenus, but as devotees of that convenient Victorian moral shibboleth, Mammon.

Dr Johnson, in his Dictionary of 1755, both echoed and confirmed this literary tradition by defining a stockjobber as 'a low wretch who gets money by buying and selling shares in the funds', and it needs to be mentioned that some writers on commercial matters would have agreed with him. M. Postlethwayt's *Dictionary of Commerce* (4th edn., 1774), depicts the jobber as little more than a parasite. An inveterate enemy of jobbing, Thomas Mortimer, wrote *Every Man his Own Broker* in 1761, and nearly fifty years later, in 1810, he published *The Nefarious Practice of Stockjobbing Unveiled.*[11] Both works are overstatements of a case, but there is more than a grain of truth in Mortimer's strictures, since jobbing was inherently open to abuse.

Although the jobber, as distinct from the broker, was notionally a wealthy man, his wealth was in many cases illusory, not necessarily with intent to defraud, but because of the very nature of stockjobbing.

[10] The South Sea Company was reconstituted on a more sober basis soon after the crash of 1720. It continued to trade, with a monopoly in its sphere of operation, until 1807. By 'South-Sea Broker' Swift could be referring to a jobber who had made a fortune through the Bubble and retired unscathed; it is possible, though, that he is thinking of one of the directors of the restored company.

[11] Ingrained distaste for the profession informs the whole style of Fox Bourne's description. (See especially Ch. 11 of his *The Romance of Trade.* See also C. Duguid, *The Story of the Stock Exchange* (1910), pp. 48–50.)

At certain periods, when there were a great number of enterprises afloat, it was always possible for the jobber to 'buy' or 'sell' great numbers of shares or quantities of stock, which he neither possessed nor had the means to purchase. Although the Stock Exchange was open daily, the ownership of stocks and shares could be transferred only at specified times, known as delivery days. Thus it was possible to gamble by purchasing shares in a company, say, a fortnight before delivery day, hoping that their value on the market would rise—or acquiring secret information to ensure this—and then selling them immediately on the day, pocketing the difference between buying and selling prices as a clear profit. Such a man, speculating for a rise, is known as a 'bull'. A profit can also be made by reversing the process, and speculating for a fall, such a spectator being a 'bear'. Bulling and bearing are not, or need not be, dishonest practices, but they are clearly open to abuse. Throughout the eighteenth century such speculations in future rises and falls were regarded as particularly unsavoury, and that disapproval extended into the new century. Fox Bourne was unequivocal in condemning these practices: 'Another offence which, though of long standing, has lately been committed with unexampled impudence . . . consists of . . . "bulling" and "bearing".'[12]

Closely associated with bulls and bears was the 'stag', a man who applied for a new issue of shares in the hope of making a quick profit when dealings began. It appears that the term was first used in literature by Thackeray in his cartoon 'The Stags. A Drama of Today' (*Punch*, 1845, ix, p. 104). Two humble workmen, Tom Stag, 'a Retired Thimblerigger' and Jim Stagg, 'an Unfortunate Costermonger', are seen applying for railway shares, for which they could not possibly pay, using assumed aristocratic names, and citing Peel, Wellington, and Coutts's Bank as referees. The stag theme was to form the subject of several cartoons in *Punch* during the years 1845 and 1846, and other direct references are to be found in Cruikshank's *Table Book* (1845).[13]

It is virtually certain that this term provided Dickens with the name of 'Staggs's Gardens' in *Dombey and Son* (1848):

[12] Fox Bourne, *The Romance of Trade*, p. 321.

[13] The substance of this paragraph is derived from Michael Steig's excellent and informative article, '*Dombey and Son* and the Railway Panic of 1845', the *Dickensian*, vol. 67, No. 365, Sept. 1971, pp. 145–8. Professor Steig instances many further examples of the term 'stag' in periodical literature of the 1840s.

THE STAG, THE BULL, AND THE BEAR.
(A Railway Fable.)

2. Shady promoters decamp with their ill-gotten gains during the Railway Mania of the mid-forties.

> Some were of opinion that Staggs's Gardens derived its name from a deceased capitalist, one Mr. Staggs, who had built it for his delectation. Others . . . held that it dated from those rural times when the antlered herd, under the familiar denomination of Staggses, had resorted to its shady precincts. (Ch. 6.)

The conjunction of the name 'Staggs' with a 'deceased capitalist' is probably sufficient to show that Dickens knew the term, and that there may be a sort of allegorical metaphor in the expressions 'antlered herd' and 'shady precincts' to suggest the practice prevalent, during the Railway Mania, of speculating in time-bargains.[14] We find Kingsley, too, preoccupied with these activities in his novel *Yeast* (1851), where Lancelot, the idealistic hero, condemns professedly religious people of a puritan turn of mind who devote themselves to 'the Stock Exchange, and railway stagging . . . and the frantic Mammon-hunting which has been for the last fifty years the peculiar pursuit of the majority of Quakers, Dissenters, and Religious Churchmen.'[15] It is interesting to notice here how what had been originally the province of the jobber has spread to the independent investor, and that Mammon-hunting has infected even the righteous!

A novelist of the calibre of Trollope could take these practices and develop from them an accurate and credible exposé of the financial mores of his time, as he does in his masterpiece *The Way We Live Now* (1875). Others, though, used jobbing and jobbers for purposes of creating melodramatic villainy, perpetuating the 'low wretch' image with little reference to reality, and with results often amusing to the modern reader.

In his novel *Frank Fairlegh* (1854), Frank Smedley created a splendid Victorian villain, Richard Cumberland. As a youth, he is addicted to 'the billiard table', and in order to pay his debts he is obliged to

[14] See Steig, ibid., p. 146. Professor Steig suggests that the passage quoted is a conscious allusion to stagging, but does not examine the expressions that I have isolated for comment. 'Antlered herd' could well suggest a crowd of stagging speculators, and 'shady precincts' could connote, albeit indirectly, the many shady offices of the 'alley men', crooked jobbers, who conducted their business in the alleys surrounding Capel Court, where the Stock Exchange stood. See, for instance, Richard Doyle's cartoon, ' "Stag" Stalking in Capel Court' in *Punch* (ix. 172 (1845)), where the antlered speculators could well be seen as an 'antlered herd'.

[15] *Yeast*, Ch. 2 (Eversley Edition, 1902). Serialized in *Fraser's Magazine*, July–Dec. 1848, the date of the action is given in Ch. 14 as 1849. It appeared revised in book form in 1851. Kingsley's outburst was probably occasioned by the commercial crisis of 1847–8, which arose from the Railway Mania and the food and money panic of that time.

borrow 'five-and-twenty pounds' from his tailor. Entirely without moral scruple, he utters his *credo* in Chapter Four: 'the world is made up of knaves and fools—those who cheat, and those who are cheated—and I, for one, have no taste for being a fool.'[16]

In desperate straits, Cumberland attempts to embezzle a cheque for £300. He is detected, and expelled from Dr Mildman's school. Some years later he re-enters the story, this time as the blackmailer of an even greater villain, Welford. Amazed that Cumberland has developed into such a villain, Frank asks his friend Freddy, a lawyer, for details, and receives the following reply:

> Cumberland . . . has become somehow connected with a lot of bill-brokers—low stockjobbers—in fact, a very shady set of people, with whom, however, in our profession, we cannot avoid being sometimes brought into contact; he appears, indeed, himself to be a sort of cross between blackleg[17] and moneylender . . . the thorough and complete blackguard.' (Ch. 26.)

One notices how jobbers and brokers are lumped together as being 'low' and 'very shady', and how the term 'moneylender' is brought into close proximity with them. The suggestion that financiers are 'moneylenders' is common in many Victorian novels, and it may be added here that a moneylender was usually a villainous Jew, exercising one of the two professions thought suitable for that race, namely, usury, or selling old clothes. Although apparently part moneylender himself, Cumberland is about to be arrested for a debt of £750 to another usurer. He flees to America, takes to drink, and finishes in 'abject poverty'.

The tradition of jobbers as vulgar upstarts is well typified by Mr Copperas, a character in Bulwer's *The Disowned* (1829), whose depiction reveals a great deal about literary prejudice against operators in the money market.

Copperas is no dissipated swindler, but he does represent, we learn, 'the respectable coxcombry of the counting-house, or the till' (I, Ch. 7, p. 89).[18] He lives in a small, pleasant suburban house, and is honest,

[16] Compare Jonas Chuzzlewit's rule of life: 'Here's the rule for bargains. "Do other men, for they would do you." That's the true business precept. All others are counterfeits.' (*Martin Chuzzlewit*, Ch. 11.)

[17] The term 'blackleg' is used here in its older sense of 'swindler', in particular one associated with 'the turf'.

[18] E. L. Bulwer, *The Disowned* (2nd edn., 3 vols., 1829). Page references in the text are to this edition. (Throughout the present work, this author is cited as Bulwer. This avoids yet another recounting of his manifold names, styles, and titles.)

witty, and not particularly rich. His wife, a lady with social pretensions, has a shelf containing 'all the best English classics', including Young's *Night Thoughts* and Pope's *Iliad*.

Living as a lodger in this suburban villa, the hero, Clarence, meditates:

> this base, pretending, noisy, scarlet vulgarity of the middle ranks, which has all the rudeness of its inferiors, with all the arrogance and heartlessness of its betters — this pounds and pence patchwork of the worst and most tawdry shreds and rags of manners, is alike sickening to one's love of human nature, and one's refinement of taste . . .' (I. Ch. 11, pp. 136–7.)

These are strong words; and although Mr Copperas plays no important part in the story, Bulwer contrives an incident in which he can be revealed in his true, craven stockjobbing colours. His neighbour, the rich and noble Mr Talbot, is threatened by burglars, and young Clarence begs the stockjobber to lend him his pistols in order to effect a rescue.

> 'I shall commit no such folly,' said the stockjobber; 'if you are murdered, I may have to answer it to your friends, and pay for your burial. Besides, you owe us for your lodgings . . .' (I. Ch. 18, p. 212.)

Clearly, we must not expect taste and noble refinement from such representatives of the 'middle ranks', and certainly not from 'low stockjobbers'!

How far do these fictions accord with reality? By the 1840s the profession of stockjobbing had become not only respectable, but essential to the operation of both public and private commerce. Despite opposition from Government and from City authorities,[19] the jobbers and brokers slowly but surely established themselves as a responsible and important financial body as the eighteenth century advanced. In 1762 the principal stock dealers formed a confederation or club that met at Jonathan's Coffee House in 'Change Alley. In 1773 they moved to new premises in Sweeting's Alley, which they now called the 'Stock Exchange'.

[19] Supposedly, jobbers and brokers had to be licensed, under statute of William III, by the Lord Mayor and Aldermen, thus becoming legal 'exchange brokers', able to ply their business, together with merchants and others, at the Royal Exchange. In practice, however, their numbers and importance were too great for this archaism to have any meaning, and legislation to limit their practice was always ineffective. For the early legal standing of stockjobbers and brokers see in particular Duguid, pp. 56–7.

"STAG" STALKING IN CAPEL COURT.

3. Mr Punch prepares to drive out the railway 'stags' from the Stock Exchange. 'Stagging' was much abused, and often condemned.

The last two decades of the century provided these organized brokers with considerable business in raising war-loans and financing the expanding canal system, so that when in 1802 they moved into their own new Stock Exchange in Capel Court, their essential role in public and private finance was firmly established. To belong to the Stock Exchange after 1800 was to advertise yourself as a man of integrity: less reputable jobbers and brokers were not admitted to 'the House', but plied their trade in the adjacent streets and alleys.

Anyone who attempted to form a view of stockjobbing from reading Bulwer or Smedley would thus be seriously misled. The Dutch scholar H. Jansonius, in his thesis *Some Aspects of Business Life in Early Victorian Fiction* (1926), observed of this tradition:

> The popular idea of the work of 'the gentry of 'Change Alley and the coffee houses', as represented in literature, seems to have been that they went on 'Change in the morning to speculate there, and either made a fortune, or went into the Gazette at one stroke. (p. 131.)

Thus in T. L. Peacock's *Crotchet Castle* (1831), with reference to jobbing speculation during the 1825 company mania, we learn that young Mr Crotchet

> had made himself a junior partner in the eminent loan-jobbing firm of Catchflat and Company. Here, in the days of paper prosperity, he applied his science-illumined genius to the blowing of bubbles, the bursting of which sent many a poor devil to the jail, the workhouse, or the bottom of the river, but left young Crotchet rolling in riches. (Ch. 1.)

The reader will be on his guard against confusing these literary exaggerations with the realities of stockjobbing and its practitioners; he needs to be careful, too, when reading works by authors who seem well informed about the facts of commercial life. For instance, in Samuel Warren's *Passages from the Diary of a Late Physician* (1838),[20] Mr Dudleigh, a merchant, amasses a large fortune by establishing a monopoly in nutmegs, and Warren explains in accurate detail how this form of commodity speculation could be carried on with great profit (pp. 231–2). The reader is impressed with this convincing detail. However, once Warren involves Dudleigh in finance *per se*, he becomes uncertain of procedure. Dudleigh, we are told, 'made one of the most

[20] The stories appeared in *Blackwood's Magazine*, Aug. 1830–Aug. 1837. They appeared in book form at intervals, the third and final volume in 1838. Quotations in the text are from Warren, *Works* Vol. i (Blackwood 1854).

fortunate speculations in the funds which had been heard of for years', thus bringing his fortune to 'a quarter of a million' (p. 231). 'The funds', with their modest yield, seem out of place here as a 'speculation', a word in this context connoting a bold and audacious gamble. We are soon in familiar territory when we learn that 'The brokers . . . came about him, and he leagued with them', venturing successfully into bill-discounting. 'Then again, he negotiated bills on a large scale, and at tremendous discounts; and in a word, by these and similar means, amassed, in a few years, the enormous sum of half a million of money' (p. 232).

No bill-broker could have made his fortune in this way. If 'tremendous discounts' were offered on bills of exchange, then such bills would have been near their falling-in date, and thus low in value. Such bills, known as 'third-rate paper', no matter in what quantities bought up, are not the stuff on which fortunes are founded. Warren's purpose here is to show how a man sets out to amass an enormous fortune which he will later lose in a crash. To produce this fall, Warren detailed every means he knew, or thought he knew, to build up Dudleigh's fortune in a few years, so that he could lose it all in circumstances conducive of moral commentary.[21]

Warren was strong on mercantile practice, but weak on pure finance. The reverse is true of Dickens, whose merchants, such as Mr Dombey or Anthony Chuzzlewit, are vaguely realized, but who, in *The Pickwick Papers* (1836–7), shows a remarkably accurate understanding of Stock Exchange practice. When old Mr Weller's wife dies, he determines to invest £200 of her estate in

> 'them things as is alvays a goin' up and down, in the City . . . Them things as is alvays a fluctooatin', and gettin' theirselves inwolved somehow or another vith the national debt, and the checquers bills, and all that.'
> 'Oh! the funds,' said Sam.
> 'Ah!' rejoined Mr. Weller, 'the funs; two hundred pounds o' the money is to be inwested for you, Samivel, in the funs; four and a half per cent. reduced counsels, Sammy.' (Ch. 52.)[22]

[21] A fuller account of Dudleigh's peculiar business practices before his madness will be found in Jansonius, *Some Aspects of Business Life* pp. 95–6. Jansonius remarks that the ways in which this 'king on 'Change' tries to procure ready money at a time of crisis are so strange 'that the reader is inclined to ask whether his senses have not entirely deserted him'.

[22] 'Two hundred pounds vurth o' counsels to my son-in-law Samivel, and all the rest of my property . . . to my husband, Mr. Tony Veller . . .' (*The Pickwick Papers*, Ch. 55). As this transaction takes place after August 1830 (the date of the writ against Mr Pickwick given in Ch. 20), it is possible that Mrs Weller's holding was in the issue of 1827.

The 'funds', frequently mentioned in nineteenth-century novels, were always a safe investment, and until the outbreak of the Great War in 1914 called for new ways of raising government loans, the 'funds' were synonymous with 'Consols', Mr Weller's 'counsels'. The word is an abbreviation of Consolidated Annuities, a term used after 1751 to apply to the consolidation in that year of the major public loans, raised in the form of redeemable or perpetual annuities paying interest at the rate of 3 per cent. It later came to be applied to government stock issues, so that by 1815, for instance, there was in existence £800 million of Consols, paying 3 per cent. Mr Weller's reduced Consols, which are paying four and a half, were, in fact, reduced annuities. By 1815 the National Debt stood at £861 million, paying £27 million annually in interest. Many funds were consequently reduced by Acts of 1824, 1827, and 1834, and it is in such a fund that Mr Weller decides to invest his £200.[23]

Consols were transferred not by deed, as were other stocks and shares, but by entries in registers kept in the Bank of England, which was also responsible for paying the interest. Mr Weller is accordingly introduced to a stockbroker, the celebrated Wilkins Flasher, Esq. There is nothing at all 'shady' about this volatile character, whose surname, however, recalls the 'scarlet vulgarity of the middle ranks' so deplored by Bulwer.

Flasher conducts Mr Weller and his friends to the Bank of England, where they enter the Consols Office. From the mid-eighteenth century the Consols market had gradually removed from the Bank to the Stock Exchange, though until 1834 dealings in the funds could, and did, take place in the Rotunda of the Bank. The Bank Act of 1834 enabled the Bank to rid itself of this nuisance. The description of the Consols Office, and its counters with their alphabetical sections, is true to reality, as is the means whereby Mr Weller signs the transfer of his wife's holding to himself as executor.

Mr Weller wishes to sell his portion of the funds, so the party repair to the Stock Exchange to effect the sale. Dickens says that they 'proceeded to the gate of the Stock Exchange', and that Mr Flasher, but not Mr Weller, entered the building. This again is accurate,

[23] Consols paid 3 per cent until 1887, when they were reduced to the present rate of 2½. Although the interest on Consols is low, the purchaser gains from their being sold at a discount. In 1815, the average price was £65 per cent. Throughout the century they rose in value, reaching par, and then, in 1898, over £110 per cent. They fell gradually from that time.

as Mr Flasher, who is a member, may enter the Exchange, but Mr Weller may not. Flasher has no trouble selling Mr Weller's holding, and, 'after a short absence, returned with a cheque on Smith, Payne, and Smith, for five hundred and thirty pounds . . . and Wilkins Flasher, Esquire, having been paid his commission, dropped the money carelessly into his coat pocket, and lounged back to his office.' (Ch. 55.)[24]

There is something reassuringly sane and reasonable about this whole episode, which can deal with the Stock Exchange without bloated rhetoric and moral sermons; it is an accurate glimpse of what actually went on in the City, and how the majority of 'low wretches' actually conducted themselves. Even the bank chosen by Dickens to encash the cheque has an impeccable pedigree. Originally a Nottingham house, Smith, Payne and Smith established itself in London in 1758. It never crashed, or did anything to warrant an appearance in Bulwer or the *Gazette*, and, via the National Provincial Bank with which it merged, is an ancestor of the present National Westminster Bank. In Mr Weller's day, the clearing establishment for those London banks who used the clearing system was housed in Smith's premises.

Not all novelists adopted the traditionally hostile view typified by Bulwer's and Smedley's creations. Disraeli, not only aware of the leading role of the Stock Exchange in the country's financial enterprise, but also a personal friend of Baron Lionel de Rothschild, exalted the figure of a stockjobber to virtually superhuman heights when he created Sidonia, the theocratic financier, in *Coningsby* (1844) and *Tancred* (1847). Charles Kingsley's Lord Minchampstead in *Yeast* (1851), is also a jobber, but one whom Kingsley admires for his power to benefit others, and for his personal qualities. Minchampstead is an accurate presentation of a type frequently encountered as a leader of Victorian finance, the entrepreneur with interests in many and varied undertakings, embracing manufactures, banking, jobbing, and land-purchase: 'From a mill-owner he grew to coal-owner, ship-owner, banker, railway director, money-lender to kings and princes; and, last of all . . . landowner.'

[24] Dickens omits one minor transaction. Before 1911, it was essential for the seller to attend personally at the Bank of England to sign the transfer. After their visit to the Stock Exchange, therefore, old Mr Weller would have returned to the Consols Office to authorize Sam's holding of £200.

Minchampstead fulfils no active role in the novel. He is merely described in Chapter Six as a type of self-made giant of commerce whom Kingsley is able to admire and respect. 'Moneylender to kings and princes' is a favourite contemporary expression for jobber, implying, with some justification, that these men could directly influence the policies of government. Kingsley wishes us to understand that Minchampstead is on a par with Rothschild as a financier of foreign loans. We notice an echo of the hostile tradition in the word 'moneylender', with its special connotations, but it is tempered here with a sort of distantly-observed awe.

Mrs Gore, whose novels show a serious awareness of the distorting influence of the anti-jobber tradition, realistically defends the practice of the Stock Exchange in her novel *The Moneylender* (1843). In this work an artist, Verelst, though considering himself to be a victim of the stockbroker Osalez, has no illusions about the importance of Exchange business to the prosperity of the nation: 'I had always fancied that Exchange speculators, so long as prosperous, occupied an important position in the moneyed world! . . . Without them, how can the finances of kingdoms be carried out?' (II, Ch. 6.) It is interesting, too, to see how another character in the same novel draws a sharp distinction between stockbrokers, whom he calls 'the first financiers and most respected men in the country', and a moneylender, who is 'a common usurer . . . a Jew . . . a miser . . . an extortioner . . .' (II, Ch. 6). Here, at least, is one novelist who emphasizes a distinction that writers such as Bulwer and Smedley were content to blur.

'Stockjobber', and the equally inelegant term 'loanmonger', have come to have a pejorative ring, and this is due in great measure to popular literary use of the terms. They may both be quite accurately applied to such distinguished financiers as Nathan Meyer Rothschild and his son Baron Lionel de Rothschild, Sir Francis Baring, and many others who played crucial roles in private, national, and international finance. Fox Bourne can use one of these terms in its neutral connotation when he describes the elder Rothschild as 'the very type and perfection of a stockjobber', and, despite his strictures on abuses, he views jobbing with sober realism when he writes: 'This trade, within proper limits, is, of course, a necessary branch of English commerce; and for many generations it has been honestly pursued by honest men without number.'[25]

[25] Fox Bourne, *The Romance of Trade* pp. 321, 323.

There are many dangers inherent in the practice of seeking the facts of Victorian commerce in the fiction of the period. Of course, many of the facts are there, but they are often coloured by literary prejudice, by the demands of a story, or by a novelist's response to what he thinks his readers would appreciate and enjoy at the time of writing. Stockjobbers and brokers were particularly vulnerable to parody and exaggeration, for the reasons shown, though some novelists, notably Mrs Gore, take pains to rectify some of these popular distortions. One hopes that the reader, when next happening upon a financial character in a nineteenth-century novel, will exercise caution before assuming that what he reads necessarily reflects total reality.

2

Crisis and Mania

The commercial crisis of 1825, and the period of wild and widespread speculation preceding it, created a lasting impression in the minds of business men, their chroniclers, and fictionalizers. The following account, detailing the progress of this 'company mania' and its aftermath, shows the 'economic rhythm' in action, and incidentally reveals the dramatic realities of these commercial convulsions in contrast to the sober abstractions of the political economists.

The year 1824 saw the development of a mood of financial optimism and a desire for new enterprises, not only in entrepreneurs but in the investing public at large. The stringencies of the war period had been all but dissipated, money was plentiful, and interest rates low. The Funds were returning lower rates than previously, and holders began to look for new and speculative channels of investment. It happened that such a new and apparently irresistible source of investment income lay in the taking-up of South American government loans.[1]

Britain had recognized the independence of Colombia, Mexico, and Buenos Aires, and would soon extend recognition to Peru, Chile, and Guatemala. Government confidence in these fledgeling states influenced public opinion, and a great eagerness developed to invest surplus capital in the national loans of these countries. This was the first stage in what Fox Bourne called 'the greatest English mania since the South Sea Bubble'.[2] It seemed to matter little whether these states were stable politically or economically. Contractors found that they could scarcely find sufficient scrip to satisfy the mania for foreign investment, and this fever continued unabated into 1825.[3]

[1] There had been sound precedent for investing in foreign loans: European governments, re-establishing their credit after the convulsions of the Napoleonic era, had arranged for the floating of loans in London, mainly between 1817 and 1823, and these yielded high return, with dividends regularly paid.

[2] *The Romance of Trade*, Ch. 11: '. . . thus was opened up to England that same trade with the eastern shores of the Pacific from which the South Sea Company had expected to derive boundless wealth' (loc. cit.).

[3] For instance, the Peruvian loan of 1825 was for £616,000, and a sum of £480,000 was quickly raised. This was from instalments purchased, not from paid-up contributions. Speculators seemed unaware that they were liable for the full amount of their

Closely connected with this mania for foreign loans was the floatation in London of numerous companies to work the silver mines of Mexico, Peru, Chile, and Brazil, which had been temporarily abandoned during the civil wars. In the public mind these companies afforded glittering prospects of wealth, and investors flocked to the brokers to invest their capital in ventures of every kind connected with the mines and their exploitation.[4]

This sudden burst of frenzied investment in foreign loans and mining enterprises brought in its wake a wave of hundreds of paper companies in the home market: the days of the South Sea Bubble seemed to have returned a century later. Before the passing of the Company Act of 1825, joint-stock companies were virtually obliged to operate independently of the law, so that no investor in such a company could seek legal redress for the grossest abuse of trust.[5] This in no way stemmed the tide of investment in the hundreds of companies that now appeared, luring prospective clients with prospectuses containing ridiculous, imprudent, and often dishonest proposals.[6]

During 1824 and 1825, no fewer than 624 company schemes were floated. Of these, 143 died almost at birth; 236 issued prospectuses,

subscriptions in any loan or company, and this obvious truth had to be stated publicly in Parliament (*Hansard*, xiii, 1032.) Similar fevered trafficking took place for loans to Brazil (1824, £1 million; 1825, £2 million), Mexico (1825, £3,200,000), Naples (£2½ million), and Greece (£2 million). All were massively subscribed.

[4] The great temptation offered to the speculator is clearly explained in the *Annual Register* for 1825 (3):

> 'In all these speculations, only a small instalment, seldom exceeding five per cent, was paid at first, so that a very moderate rise in the price of the shares produced a large profit on the sum actually invested. If, for instance, shares of £100, on which £5 had been paid, rose to a premium of £40, this yielded on every share a profit equal to eight times the amount of money which had been paid. This possibility of enormous profit, by risking so small a sum, was a bait too tempting to be resisted.'

[5] The imperfect state of company law in the early decades of the nineteenth century is explained in the Introduction. Disraeli, in *Lawyers and Legislators* (1825), p. 95, declared that Government ought 'to make joint-stock companies amenable to the law of which under the present system they are forced to be independent'. Immediately after the crash, an Act (1825: 6 Geo. IV, c. 91) made joint-stock companies amenable to the common law. This imperfect move had little effect on commercial life, and in any case, the damage had been done.

[6] Fox Bourne rightly points out that not all companies floated in 1824–5 were unsound. Many survived, and provided legitimate services (*Romance of Trade* Ch. 11). Among the survivors may be mentioned the Liverpool–Manchester Railway, the General Steam Navigation Company, and forty-four mining companies. The Alliance Fire and Life Assurance Company, founded by Nathan Meyer Rothschild in 1824 with a capital of £5 million, is, of course, still in business.

but never offered shares; 118 opened markets in their shares, but were later abandoned. One hundred and twenty-seven survived into 1827.[7] Prospective investors, when the rage was at its height, flocked to put their money into undertakings that were ill-advised, fraudulent, or ridiculous. There was one company, which actually found subscribers, that proposed to drain the Red Sea in order to recover the lost treasures of Pharaoh and his host.[8]

Throughout 1825 the mercantile world moved towards the inevitable crisis.[9] Trading of all kinds continued unabated, encouraged by the ease with which accommodation was made available. Both the Bank of England and the country banks were eager to advance money cheaply to finance all manner of enterprises, and the number of bank notes in circulation increased dramatically. There was eight million pounds more paper money in circulation in 1825 than in 1823, with no corresponding increase in trade and industry to justify it. At the same time there had developed a vast extension of private credit—the 'new currency' of the age—and the market was flooded with bills of exchange, promissory notes, and similar paper. This facility had caused merchants and others to begin speculating in basic commodities, thus effecting a rise in prices, and a consequent increase in imports. Meanwhile, the foreign loan and mining investments had to be met.

The massive speculations, and the eagerness with which money was loaned by the banks, caused paper sums gradually to overtake the

[7] Figures from H. English, *A Complete View of the Joint-Stock Companies* . . . (1827), p. 28. By the opening months of 1825, 276 of these companies had already been projected. Morier Evans, in *The Commercial Crisis 1857–1858* (1859), pp. 14–15, gives an interesting break-down of these 276 companies: canals and docks (33); railways (48); gas (42); milk (6); water-supply (8); coal-mines (4); metal-mines (34); sea-water baths (2); insurance (20); banking (23); navigation and packets (12); fisheries (3); newspapers (2); Thames tunnels (2); 'embellishment of London' (3); miscellaneous (34). Evans states that the aggregate paper capital of these companies was over £174 million, of which it appears that only £17½ million was actually called up during 1825. (Smart, *Economic Annals*, ii, p. 295.)

[8] J. A. Francis, in his *History of the Bank of England*, ii, Ch. 3, states that he actually saw the prospectus of this company. Morier Evans remarked that 'the very names of the "Bubble Companies", as they were called, if now quoted, would look like a sarcasm upon speculation in general' (*Commercial Crisis 1857*, p. 1). There are many reflections of pompous and ridiculous names of such companies in nineteenth-century fiction, and some of these will be found in succeeding chapters of the present work.

[9] An article in the *European Magazine* for April 1824, declared: 'Never since the South Sea Bubble has the mania been so endemic. There is not a capitalist nor moneylender all over the empire that is not infected with it, and where it will end, no man can foresee.' In fact, as early as April 1824 the end of this mania could clearly have been foreseen, though it would not, perhaps, have been prudent at the time to say so.

amount of real capital available in the country. In November 1824 the Bank bullion reserve was £11,323,760. Towards the end of November 1825, it was down to £3,012,150. Alarmed, the Bank quite suddenly diminished its issues, and raised its rate of discount. Many of the bubble companies were failing, and public suspicion came to include the banks, many of which, following the Bank of England's example, began to retrench their paper issues and to refuse accommodation. Inevitably there was a run on the banks, which lasted into the new year. Many banks survived by resorting to quite desperate measures. A bank at Oxford piled up all its gold reserve in dishes on the counter, and reassured clients made no demand for cash redemption. The famous Norwich bank of Gurney's stopped a run by displaying on its counter a pile of Bank of England notes several feet thick.

On 29 November 1825, the great provincial bank of Elford's at Plymouth stopped payment, and a few days later the Yorkshire house of Wentworth & Co. closed its doors. These two failures precipitated the crisis of 1825. The effect of the closures was immediately felt in London, and on Saturday, 3 December, the old-established London bank of Sir Peter Poole, Thornton & Co. suspended payment. Poole's, in common with many other banks, had failed to ensure that it had reserves sufficient to meet at least part of its paper liabilities. It was agent for a large number of country banks, and, despite the perilous state of its own bullion reserves, the Bank of England rushed to the support of this important house. Its efforts were in vain, and Poole's collapsed on Monday, 12 December. Panic increased, and frantic crowds of depositors appeared in Lombard Street. A number of important London banks failed within the week, ruining those country banks that were dependent on them. During the period 1825–6 some eighty country banks were broken, leaving behind them a trail of misery and destitution.[10]

The frenzied scramble for gold was now directed at the Bank of England, and by 24 December 1825, the coin in the Bank was reduced to £426,000, and the bullion to £601,000; a total of £1,027,000.

[10] The number of banks said to have failed in this period varies in different authorities. Clapham, *Economic History*, I, p. 273, footnote 2, cites 80, 36, 76, 63, 79 from different sources. Morier Evans, *Commercial Crisis 1857*, p. 2 (footnote), gives the number for 1825 as 79, with 58 branches, and for 1826, 25 with 2 branches. Total liabilities for these banks he computes at £4,608,000. A considerable number of these broken houses were ultimately able to pay 20*s* in the £, and to resume business.

Fortunately, the public still reposed confidence in the Bank's paper money, and what many saw as a near-miracle prevented the total collapse of the country's monetary system. A box containing 700,000 one-pound notes was discovered in the Bank cellars, having lain forgotten for a year. The issue of these notes, combined with a public statement of confidence from the London merchants, was effective in turning the tide. By 26 December the panic was over, but business men would never forget how near the country had been to a state of barter.[11]

Although by the end of December the Bank was safe, and a more sober, responsible credit had revived, the disastrous results of the speculation mania continued into the new year. In November 1825, 188 enterprises had gone bankrupt, and the number rose to 220 in December, and to 321 in January 1826. This figure increased to 380 in February, and in March 315 firms failed, 93 of them on a single day. Most of these failed enterprises were the bubble companies that staggered on for a few weeks after the money crisis of December; thousands of investors were beggared by their demise.

They were joined by the avid speculators in South American mining companies, nearly all of which merely absorbed the capital sent out of the country, and then closed down. Attendant schemes connected with the mining operations were simply abandoned. Those who had placed their hopes in South American loans fared no better. Over £11 million was paid on foreign loans in 1825.[12] This money, so eagerly exported to the New World, was used to finance the revolutions and civil wars that characterized the infant states. Little or no interest materialized, and this, combined with the absence of security, and the rise in the home rate of interest, caused the stocks to be thrown on to a depressed market where, in effect, they became merely waste paper.[13]

[11] The famous one-pound notes quickly circulated to solvent banks that anticipated dangerous runs from the public. It was possibly a pile of these notes that Gurney's displayed at their bank in Norwich. The Bank was further able to increase its reserve by negotiating a loan from the Bank of France of £2 million, repayable in three months' bills drawn on London.

[12] *Annual Register* (1825), p. 48 (footnote).

[13] Smart, *Economic Annals*, ii, pp. 294–5. The suffering of individuals was considerable. The South American loans entailed the loss of nearly all the sums subscribed. Only a few small dividends were paid, and these came out of the subscribed capital. (See Tooke, *History of Prices*, ii, p. 159.)

There were other commercial crises in the nineteenth century, but none so devastating, dangerous, and alarming in the public mind as this speculative convulsion. It was to be remembered throughout the century with the same intensity that its famous ancestor, the South Sea Bubble, had created in the minds of those who had either lived through it, or heard of it as part of national folklore. Victorian economists and chroniclers of business would in their own writings hark back to that period, which to many was still within living memory, and describe its progress in a style very similar to that of the novelist. A dramatic, inflated tone is often employed, and one realizes that such a style was not necessarily the prerogative of the novelist. Fox Bourne deals in realities which are in essence as heart-rending as any reflections in the fiction of the day:

> Old Londoners still remember with a shudder . . . how hundreds of men, supposed to be rolling in wealth, wandered distractedly up and down the streets, fearing that in an hour or two they might be bankrupts; and how thousands watched their movements with ghastly interest, knowing that the bankruptcy of those in whom they trusted meant starvation to themselves. Many did starve . . . Old men lay down to die under the load of disgrace that they had brought on themselves . . . and the silent wretchedness of widows and orphans, robbed of their all, furnished a strange contrast to the gratulations of those who had prompted the mania and made money out of the panic. (*The Romance of Trade*, Ch. 11.)

Literary reflections of the 1825 mania began early, with the appearance in 1828 of Benjamin Disraeli's *The Voyage of Captain Popanilla*. The young Disraeli's factual interest in the crisis had been reflected during its progress in two analytical works: *An Enquiry into . . . the American Mining Companies* and *Lawyers and Legislators: or Notes on the American Mining Companies*, both of which appeared in 1825. Two years later, his creative genius used the events of the mania as the basis of a highly effective politico-economic satire.

The story of *Captain Popanilla* begins in a tropical utopia, uncorrupted by European institutions, the island of Fantaisie, where a young man called Popanilla finds a library of 'useful knowledge' cast up from a wreck. After reading the books he is converted to Utilitarianism:

> Popanilla, therefore, spoke of man in a savage state, the origin of society, and the elements of the social compact, in sentences that would not have disgraced the mellifluous pen of Bentham . . . If there were no utility in pleasure, it

was quite clear that pleasure could profit no-one . . . consequently pleasure is not pleasant.[14]

Exiled for his novel ideas, Popanilla sails to the Island of Vraibleusia [Britain], where he makes his way to the capital, Hubbabub. He is befriended by a Mr Skindeep, who occasions some paradoxical images of Benthamite theory and social practice. A beggar, asking for alms, is offered money by Popanilla; his mentor checks his 'unphilosophical facility' (p. 390), and advises the beggar to work. Two merchants, anxious to sell Popanilla a purse, undercut each other so keenly that one of them is virtually obliged to give him the purse gratis: 'This is not Cheatery; this is Competition!' (p. 392.)

Popanilla accompanies Skindeep to his bank, where he is introduced to the intricacies of 'the Great Shell Question' (p. 397). The Government has a monopoly of pink shells, and the organs of commerce use other shells to conduct their business. The pink shells represent the Bank's notes—only the Bank of England was permitted to issue promissory notes in the capital. The other shells mentioned are the country banks' notes, and the increasing numbers of bills of exchange and other paper that were rapidly making credit the currency of the age. In the context of the 1825 mania, there is a particularly relevant dimension of irony in Disraeli's wry assertion that 'it was evident that the nation who contrived to be the richest people in the world while they were over head and ears in debt must be fast approaching to a state of perfection.' (p. 399.)

It is not long before Disraeli introduces into his work a reflection of the initial preoccupations of the year 1824; the taking-up of South American loans, and the rush to invest in the gold and silver mines. In *Popanilla* the two activities are combined when the 'Emperor of the East' offers his 'unequalled' gold and silver mines as security for loans from more wealthy nations.[15] The reaction of the Vraibleusian government to this seductive offer parallels that of Britain and the Bank during the mania: 'The cause was so reasonable, and the security so satisfactory, that the Vraibleusian Government felt themselves authorized in shipping off immediately all the gold in the island. Pink

[14] Page refs. to *Popanilla* are from B. Disraeli, *Novels and Tales* (Hughenden edn., 1881), iv.

[15] Disraeli may have read Capt. Head's *Rough Notes* (1826), which had shown that the new South American states were sadly ill-equipped for new mining operations. In *Popanilla*, we are told that the only reason the mines of Fantaisie were unworked was that the inhabitants spent all their time eating water-melons (p. 417).

shells abounded, and the stocks were still higher.' (p. 417.) The naïve view of the stability of the new nations that had been a surprising reaction from a nation old in statecraft is echoed in the Vraibleusian view that 'Although founded only last week, they already rank in the first class of nations.' (p. 418.)

A Vraibleusian mania develops when it is decided to bring to Popanilla's homeland of Fantaisie 'the blessings of civilisation and competition' (p. 425). Manufacturers and merchants rush to make and to export all manner of goods and machinery to the hitherto peaceful and 'primitive' land. Money is, of course, plentiful, and redundant capital is invested in speculative enterprises. 'Public companies were formed for working the mines, colonizing the waste lands, and cutting the coral rocks of the Indian Isle, of all which associations Popanilla was chosen Director by acclamation.' (p. 427.)

The Vraibleusian crash is occasioned by the failure of a fleet of five hundred merchant ships to locate the island of Fantaisie. The fleet consequently returns to Hubbabub, and the markets are immediately glutted with unsold goods.[16] This glut ruins the merchants and tradesmen, and a run on the government bank commences. 'As the Emperor of the East had all the gold, the Government Bank only protected itself from failure by bayoneting its creditors . . . All the millionaires crashed . . . In a word, dismay, disorganisation, despair pervaded in all directions the wisest, the greatest, and richest nation in the world.' (p. 454.)

The Vraibleusian state is saved from utter ruin by the intervention of 'a public instructor, Mr Flummary-Flum', who is clearly modelled on Ricardo. Like Ricardo, Mr Flum explained to the Vraibleusians the meaning of wages, rent, and profits, so that they soon regained their accustomed wealth.

[16] Disraeli seems to be alluding to the results of the ill-advised commodity speculation of 1824 and 1825. Speculation in cotton, silk, wool, and flax had developed into a widespread stockpiling of generally available goods in the hope of achieving a high rise in price. Tooke remarked that 'there was hardly any article of merchandise which did not participate in the rise' (*History of Prices*, ii, p. 140). In Samuel Warren's *Diary of a Late Physician*, Mr Dudleigh, the merchant, indulges in this type of speculation, creating a monopoly in nutmegs (*Works*, Blackwood, 1854, i. p. 231). Spices had reached 200 per cent in 1824–5, a fact very likely known to Warren, who thus adds a dimension of realism to Dudleigh's mercantile pursuits. By mid-1825, prices began to fall, and stockpiling commodity merchants found that they had a glut of unsold goods on their hands. (See also Smart, *Economic Annals*, ii, p. 190.)

The importance of *Popanilla* lies in its being a fully-conceived satire occasioned by a massive commercial convulsion, and as such it may be seen in relation to the 1825 crisis as Swift's *The South Sea Project* (1721) was to the events of the South Sea Bubble. Disraeli would not again in his fiction show such interest in commercial matters.[17] He was by no means, however, the only novelist to make the mania period a subject for fiction: explicit or derivative reflections occurred in many works for upwards of forty years after the event.

We have already encountered Bulwer as a critic of commercial institutions in *The Disowned*, which appeared in 1828. It was not until 1849, when the financial world had weathered three further crises,[18] that he turned once again to the City and its ways in *The Caxtons*.

The Caxtons, one of Bulwer's 'family pictures', is a pleasant and entertaining novel, with some memorable characters. Of particular relevance among these is young Caxton's 'Uncle Jack', who is firmly rooted in memories of the mania years 1824 to 1826. Whatever Bulwer's literary preoccupations were during the period, it is clear that the events of the mania stayed in his mind, to surface, in 1849, in a particularly amusing and irrepressible example of an innocent encompassed about by a crowd of fragile Bubbles.

Uncle Jack, we are told, is 'a great speculator; but in all his speculations he never affected to think of himself,—it was always the good of his fellow-creatures that he had at heart, and in this ungrateful world fellow-creatures are not to be relied upon!' (II. 2. p. 32.)[19]

Uncle Jack Tibbets is cast as a sympathetic character who is, however, a born speculator. There is a gentle irony in Bulwer's exposition of his motives for investing in foreign loans: 'Uncle Jack had a natural leaning towards all distressed communities . . . Poles, Greeks (the last were then fighting the Turks), Mexicans, Spaniards—Uncle Jack thrust his nose into all their squabbles!' (II. 2. pp. 32–3.) The reference to the Greek war of independence sets the action in 1824, and confirms Uncle Jack's identity as a speculator of the mania period. The paragraph quoted continues in a way that shows how Jack's altruistic interest in the fledgeling states is firmly guided by his eye

[17] It will be seen from Ch. 5 of the present work how Disraeli's great financier, Sidonia, is rarely found conducting what may truly be termed 'business' of any significance.

[18] These occurred in 1832, 1836, and 1847. The crisis of 1847 is often thought of as being exclusively concerned with the 'Railway Mania'. The facts are rather different from this popular conception, as will be explained in Ch. 8.

[19] Page refs. are to *The Caxtons* (Knebworth Edition, 26 vols., 1877–8).

for business: 'whenever a nation is in misfortune, there is always a job going on! The Polish cause, the Greek cause, the Mexican cause, are necessarily mixed up with loans and subscriptions.'[20]

Bulwer does not directly involve Uncle Jack in South American mining ventures, although he tells us that he has seriously considered the opportunities offered there for British company enterprise: he was attracted by 'certain mines of Mexico', and wished them to be exploited by a 'Grand National United Britons Company'.[21] He does, however, involve him in an interesting home mining venture that echoes some of the preoccupations of the mania era.

Uncle Jack inherits a farm in Cornwall, and is convinced by the assurances of certain 'engineers and naturalists' that coal lies beneath it. On the strength of this chimera—Cornwall then, as now, being devoid of coal—he immediately establishes the 'Grand National Anti-Monopoly Coal Company', designed to counter 'the monster monopoly of the London Coal Company' (ii, 2, p. 34). Bulwer accurately reflects here one of the preoccupations of Government during the mania years: the fear that certain of the new joint-stock enterprises would create monopolies of staple materials and commodities. When an attempt was made in 1825 to float a 'Metropolitan Fish Company', and a Bill for its establishment was introduced into the House, it was vigorously denounced as monopolistic.[22] Incidentally, the young Dickens, in *Nicholas Nickleby* (1838), shows Ralph Nickleby's 'Metropolitan Muffin Company' as an attempt to create a staple monopoly, one of several details which place the action of that novel, too, in the mania years, as I have shown elsewhere.[23]

For three years Uncle Jack's company appears to flourish, and the shareholders regularly receive 20 per cent. However, all is lost when the 'eminent engineer', Giles Compass, suddenly absconds to America.

[20] Uncle Jack is quite successful in these ventures, clearing £3,000. A 'job' in this context is, of course, a piece of jobbing speculation, and there was much trafficking in foreign loan stock during the period of the mania. The Greek loan of 1825 was for £2 million, and the Mexican for £3,200,000. I can find no record of a publicly floated Polish loan during these years.

[21] ii, Ch. 2, p. 36. Capt. Head, in his *Rough Notes*, found at Buenos Aires a Churning Company, employing Scots milkmaids to produce enormous quantities of butter. The inhabitants did not eat butter (Smart, *Economic Annals* ii, p. 329).

[22] Smart, ibid., ii, p. 296.

[23] N. Russell, '*Nicholas Nickleby* and the Commercial Crisis of 1825', in the *Dickensian*, Autumn 1981, pp. 144–50.

The passage describing the fall of the company has a familiar ring to anyone who is acquainted with the details of the 1825 crisis:

> it was discovered that the mine had for more than a year run itself into a great pit of water, and that Mr. Compass had been paying the shareholders out of their own capital. My Uncle had the satisfaction of being ruined in very good company; three doctors of divinity, two county members, a Scotch lord, and an East India director were all in the same boat. (II. 2. pp. 34–5.)[24]

Bulwer completes his portrait of an all-round speculator by involving Uncle Jack, inevitably, in the founding of joint-stock companies. His 'Grand National Benevolent Clothing Company' comes to nothing, and his 'Grand National Benevolent Insurance Company', with a paper capital of £2 million, 'dissolved into thin air'. Morier Evans mentions twenty insurance companies floated in 1825. Of these, three survived; Uncle Jack's 'Grand National' was, as it were, one of the unlucky seventeen that collapsed.[25]

Bulwer's literary stance in *The Caxtons* is far less minatory than that of his much earlier castigation of 'scarlet vulgarity' in *The Disowned*. He still regards 'money-makers' as 'cold-blooded creatures', but admits that they frequently manifest 'prudence and caution' in their dealings. His Uncle Jack is a feckless, optimistic enthusiast seeking his true vocation, which he finds ultimately as a land speculator in Australia where, like Mr Micawber, he prospers exceedingly. His creator, writing in 1849, hopes that he will avoid the snares of the coming gold-rush; and the reader too hopes that the irrepressible Uncle Jack, the survivor of 1825, will keep a steady head.

It has been shown how the crisis of 1825–6 was sparked off by the sudden failure of old-established and respected banks, in both London and the provinces. There was hardly a town of any consequence that did not exhibit the effects of these closures in ruined tradespeople, annuitants, widows, and orphans, whose literary counterparts form many a pathetic catalogue in novels where bankers figure as the victims of their own greed, imprudence, or villainy. In the middle years of the century appalling events would occur in the world of banking which

[24] It is interesting to note that, while 34 metal mines were projected in 1825, only 4 coal mines were floated. It is typical that Uncle Jack, inheriting a farm in a then flourishing centre of metal mining, should have thought that he had discovered coal where no coal was.

[25] Morier Evans, *Commercial Crisis 1857*, p. 14. The survivors were Standard Life, University Life, and Crown Life. The latter was absorbed by Law Union and Rock in 1892 (H. Withers, *Pioneers of British Life Assurance* (1951) p. 104).

would place this particular branch of commerce in the forefront of literary preoccupation; but even in those years and later, it is possible to find novelists working into their fictions reflections of the long-remembered bank failures of 1825. Of particular interest in this connection are Miss Mulock's best-known novel, *John Halifax, Gentleman* (1856), and Charles Reade's *Hard Cash* (1863).

Miss Mulock was born on 20 April 1826, and it is interesting to observe a perpetuation of the crisis year in the fiction of a novelist belonging to a younger generation than that of Disraeli and Bulwer, with no personal recollection of that time. In Chapter 26 of her novel mention is first made of 'Jessop's Bank', a modest country bank established in the former dining-room of its proprietor's house. The bank's daily transactions are not at this stage mentioned, but one may assume that Jessop was typical of his kind, a man of comparatively slender resources who offered accommodation to the local merchants and manufacturers. In the fifteen-year period from 1815 to 1830, 206 country banks failed, an alarmingly high figure, but hardly surprising, as many of these bankers were inexperienced, imprudent, or naïve. It was of this class that Lord Liverpool, in the debates following the 1825 crisis, complained, when he asserted that the prevailing law allowed 'any small tradesman, a cheesemonger, a butcher, or a shoemaker, to open a country bank, while persons with a fortune sufficient to carry on the concern were not permitted to do so.'[26]

Bankers of Jessop's type had only modest capital with which to speculate, and the wise banker confined his business to offering loans from what capital was raised through the circulation of his own notes and through the modest sums placed on deposit. Most accounts were current, and commission was charged for running them; interest was sometimes, though not always, paid on deposit accounts. Country banks usually had a London agent bank, which assisted them in their second principal activity, that of buying and selling bills drawn on London. The prudent banker discounted short bills of, say, three months' sight, which could quickly be converted into cash through the London agent. It was common, however, for the less wary banker to ruin both himself and his customers, or 'clients' as they were often called, by imprudent advances on a large scale to a single firm, or through ill-advised speculation.[27]

26. Quoted in Smart, *Economic Annals*, ii, p. 350.
27. Clapham, *Economic History*, i, p. 266.

The events of Chapter 30 of Miss Mulock's novel are set in 'this terrible 1825'. John Halifax, riding in company with Lord Luxmore's steward, Mr Brown, tells him that ten bank failures have appeared in that day's *Gazette*. Brown is morose and silent, having lost everything in 'Mexican speculations'. John Halifax's success in life arises partly from his level-headed refusal to be caught up in speculation and gambling. He remarks to his friend Phineas: 'I do think . . . the country has been running mad this year after speculation. There is sure to come a panic afterwards, and indeed it seems already beginning.' (Ch. 30.) Halifax is a manufacturer, symbolizing industrial enterprise and progress; by endowing him with the benefit of her hind-sight, Miss Mulock increases his stature at the expense of the worst elements of the landed interest, typified by the decadent Lord Luxmore, and the Mammon-centred representatives of the speculative mania. It is a clever and valid device to involve Halifax intimately with the events of the crisis year in order to show how virtue can survive unscathed.

The climax of the chapter, the threatened closure of Jessop's bank, is accurately set in December 1825:

> Many yet alive remember this year — 1825 — the panic year . . . Speculations of all kinds sprang up like fungi, out of dead wood, flourished a little, and dropped away. Then came ruin, not of hundreds, but thousands. . . . This year, and this month in this year, the breaking of many established firms, especially bankers, told that the universal crash had just begun.[28]

This epitome must have aroused in her contemporary readers a whole complex of recollections that would add to her depiction of Jessop's plight a dimension of realism perhaps not readily appreciated by present-day readers.

The novelist's dramatic instincts now come into play. At a coming-of-age dinner given by John Halifax for his son Guy, Mr Jessop, a man untainted by speculation, borrows Halifax's copy of the *Gazette*, where notice of all bankruptcies is posted, and reads there of the failure of 'W——'s', his agent bank. '"W——'s" was a great London house, the favourite banking-house in our country, with which many

[28] The style and content of some passages in this chapter suggest that Miss Mulock had before her during composition one of the standard works on the crisis, from which she drew her material. A typical work was the popular *British History, Chronologically Arranged*, by John Wade (1788–1875), which began to appear in 1839. Its detailed account of the 1825 crisis is commended by Morier Evans, *Commercial Crisis 1857*, p. 14.

provincial banks, and Jessop's especially, were widely connected, and would be no-one knew how widely involved.'[29]

The old banker collapses, his eyes staring blankly, his cheek the colour of ashes. This reaction to the news is something more than mere melodrama. Such bankers faced total ruin through the closure of their agents, and were further burdened by the knowledge that many members of their small communities would be ruined with them.

On the following morning a panic-stricken crowd collects outside the bank. Poignantly silent, it waits until five minutes after the normal opening-time of ten o'clock, when the silence is broken. Miss Mulock is economical in her effects, which for that reason carry a convincing realism: 'Then a murmur arose. One or two men hammered at the door; some frightened women, jostled in the press, began to scream.' (Ch. 31.) Halifax explains to Phineas the nature of Jessop's dilemma, one typical of that confronting many a small country banker of the era:

> Jessop's bank has such a number of small depositors and issues so many small notes. He cannot cash above half of them without some notice. If there comes a run, he may have to stop payment this very day; and then, how wide the misery would spread among the poor, God knows.' (Ch. 31.)[30]

The depiction of Jessop's difficulties continues in this vein of fidelity to fact, a characteristic of Miss Mulock's depiction of commercial matters. A 'run' on a bank was a peculiarly tragic event, as it often forced closure upon an establishment that was essentially solvent. The 'run' involved a demand for redemption of notes in gold, and the banker was legally obliged to meet his clients' demands. Time was the crucial factor: if the run commenced before the banker could call in his paper assets as security for the immediate advancement of specie, then he was ruined—his promissory notes became worthless paper. This was Jessop's predicament.

[29] It is interesting to note that Miss Mulock chooses to call her London bank 'W———'s'. The initial may have been suggested by Wentworth's, the Yorkshire house whose fall contributed to the crisis of December. The crisis of 1837 was triggered by the suspension of three merchant banks, Wildes, Wiggins, and Wilson, firms of great importance, known in commercial circles as 'the three W's'.

[30] During the Parliamentary inquest on the crisis in 1826, the country bankers were made scapegoats, and were accused of bringing about the crisis by their wholesale issue of their own low-denomination promissory notes. The subject is discussed at length in Smart, *Economic Annals*, ii, Ch. 31: 'The End of the One-Pound Notes'.

he was at this moment perfectly solvent, and by calling in mortgages, etc., could meet both the accounts of the gentry who banked with him, together with all his own notes now afloat in the country, principally among the humbler ranks . . . if only both classes of customers would give him time to pay them (Ch. 31.)

John Halifax saves the day by a series of rapid moves. First, he posts on the door a notice saying that the bank will open without fail at one o'clock, and then persuades a committee of 'provincial magnates' assembled to discuss the bank's future to stay their hand until he returns from a mysterious errand.[31] Halifax returns just after one o'clock, carrying a bag:

a precious, precious bag, with the consolation — perhaps the life — of hundreds in it![32] . . . He went right up to the desk, behind which, flanked by a token array of similar canvas bags, full of gold — but nevertheless waiting in mortal fear, and as white as his own neckcloth — the old banker stood. (Ch. 31.)

The factual provenance of Mr Jessop's 'canvas bags, full of gold', will be immediately appreciated. Jessop had sought to stave off the run by displaying his stock of gold on the counter, a detail clearly founded on widely-reported subterfuges such as that of the Oxford bank which piled up all its gold in dishes on the counter, and of Gurney's at Norwich, who prevented a run by similar use of a mountain of Bank of England notes.[33] Jessop's desperate action, however, has little effect, and the run is about to commence. He is saved when the universally respected John Halifax ostentatiously deposits £5,000 in gold with Mr Jessop. This splendidly dramatic act brings the banker's crisis to a sudden and happy termination, as confidence is restored.

Charles Reade was only eleven in the crisis year, but in his novel *Hard Cash* (1863),[34] he too reverts to the events of 1825. Though the book is principally concerned with exposing the abuses of contemporary

[31] Many provincial magnates were personally interested in the country banks, and in the debates of 1826 vehemently defended them, declaring that they were not the cause of the crisis, but rather victims of an imperfect monetary system. It is both accurate and dramatically satisfying that Miss Mulock should introduce this hastily-formed committee of powerful county influences at this stage in the chapter. (See also Smart, *Economic Annals*, ii, pp. 346 f.)

[32] Phineas remarks: 'I knew, almost by intuition, what he had done — what, in one or two instances, was afterwards done by other rich and generous Englishmen, during the crisis of this year.' (Ch. 31.)

[33] Smart, ii, p. 299.

[34] As *Very Hard Cash* it appeared in *All the Year Round*, Mar.–Dec. 1863. As *Hard Cash*, it was published in three volumes the same year.

lunatic asylums, one centre of interest is the career of the eccentric banker Richard Hardie. The principal action of the novel is set in 1847, but in outlining Hardie's career Reade dwells at some length on the position of bankers during the 1825 crisis. In that year, he writes, 'it was not one bubble but a thousand':

mines by the score, and in distant lands; companies by the hundred; loans to every nation and tribe . . . Princes, Dukes, Duchesses, Bishops, Peers, Lawyers, Physicians, were seen struggling with their own footmen for a place in the Exchange. (Ch. 6, p. 105.)[35]

Richard Hardie's father, although 'good', and 'steady', is ultimately 'drawn into the vortex', but Richard, like Miss Mulock's John Halifax, is able to see the inevitable result of over-speculation.

At five-and-twenty divine what he did! He saved the bank! . . . Richard Hardie, at twenty-five, was the one to foresee the end of all these bubbles . . . His superiority was so clear, that his father resigned the helm to him, and, thanks to his ability, the bank weathered the storm, while all the other ones in the town broke or suspended their trade. (Ch. 6, p. 105.)

This deliverance from the wholesale ruin of 1825 helps to establish young Hardie's reputation, though as will be seen, he falls victim to the temptations of railway speculation in the forties, and begins his spectacular progress from malversation to fraud, theft, and insanity.[36]

We have seen how financial elements in some nineteenth-century novels derive unequivocally from the great commercial convulsion of 1825, and how accurately those events are chronicled, whether personally experienced by the authors, or inherited as part of contemporary folklore. The prevailing belief in a ten-year cycle of prosperity leading to over-trading and crisis, whatever its economic validity, ensured that commercial life was well to the fore in the preoccupations of society. The effects of the 1825 crisis were widespread, and were still actively in memory when another crisis rocked the commercial world in 1836. The mid-forties witnessed the

[35] Page refs. are to the edition of 1868. Ch. vi is misprinted as Ch. vii in this edition; the true Ch. vii follows in the correct order.

[36] Thackeray's *The Great Hoggarty Diamond* (1841) is also set specifically in the crisis years of the 1820s, 'when, as the reader may remember, there was a great mania in the city of London for establishing companies of all sorts, by which many people made pretty fortunes.' (Ch. 2.) The story, with its depiction of the Turkey-merchant, Mr Brough, a founder of bubble companies, is examined in detail in Ch. 4 of the present work.

frenzied adventures in speculation known as the 'Railway Mania'. Authors thus found constant reminders of the vicissitudes of that world of business in which so many people directly and indirectly participated. It is little wonder, therefore, that promoters, speculators, bankers, merchants, and the rest figure so frequently in novels of the period. We shall turn next to a fuller examination of nineteenth-century banking, its dangers and disasters, and what the novelists made of it in their fictions.

3

Banks and Bankers

To the twentieth-century Englishman the word 'bank' usually means joint-stock bank: he thinks of the National Westminster, Lloyd's, or Barclay's. In the early nineteenth century, however, the joint-stock bank in London was a pioneer, struggling against the public's ingrained suspicion of joint-stock enterprises, and against deliberate obstruction from established bankers and interested politicians. Legislation in 1826 had permitted the establishment of joint-stock banks, provided that they operated outside a sixty-five-mile radius of London, and kept no place of business in the capital. Unlike other joint-stock undertakings, the members were to be individually liable for the company's debts.

By 1833 there were thirty-two joint-stock banks, and the Bank Charter Act of that year enabled the London private banks to operate with joint stock if they so desired. Primarily deposit banks, they posed no threat to the Bank of England's exclusive privileges in the City as a bank of issue. However, the Act further permitted the establishment of new joint-stock banks in London, and in March 1834 the London and Westminster Bank opened for business, with the distinguished banker J. W. Gilbart as manager. The Bank of England, hitherto the only joint-stock bank in London, now had a competitor, and so decided to offer all the opposition in its power. The resultant struggle, ending in victory for the new concern, is well known in banking history, and this successful outcome provided a strong stimulus for the development of this type of banking in the provinces.[1]

[1] Private banks were single establishments, but the new joint-stock banks had from the beginning contemplated the setting-up of branches. The London and Westminster's head office was in Throgmorton Street, and there were branches in Waterloo Place, Holborn, Whitechapel, Southwark, and Oxford Street. Convenient accounts of their struggle with the Bank will be found in J. H. Clapham, *An Economic History of Modern Britain* (1926) i, p. 510 f., and in L. Levi, *History of British Commerce* (1872 and 1880). Other joint-stock banks that came into being in these years were the London Joint Stock Bank (1836); the Union Bank of London (1839); the London and County Bank (1839); and the Commercial Bank of London (1840). The tendency to amalgamate began early, rapidly reducing the number of separate houses.

The character of these new joint-stock banks made it difficult for them to fulfil any useful literary purpose. They were 'open' enterprises, subject to regular, independent, public auditing, and largely free from the nepotism of their private competitors. The fact that shareholders were individually liable for the firm's debts ensured a corporate interest in the efficient running of the business, an efficiency that was enhanced by the character of the professional branch managers. There was little that the novelist could make out of such enterprises. They lacked the absolute control of one man, with his potential for guilty secrecy, and the possibility of 'all being revealed' in a breathless final chapter. Thus joint-stock banks are not to be found among the stuff of dramatic or sensational fiction.[2]

The same may be said of the savings banks, encouraged by Parliament in the 1780s to foster thrift among the poorer classes, whose modest savings were invested by boards of honorary trustees, usually in 4 per cent Government stock. They also received the funds of the friendly societies, which acted as depositors for their members. Extremely popular and excellently managed, there were over four hundred at the beginning of the forties, with total deposits in excess of £14 million. When they do occur in novels of the period, they are portrayed as victims of unscrupulous trustees. Mrs Gore's Richard Hamlyn, the fraudulent banker in her novel *The Banker's Wife* (1843), was such a trustee, and after his fall a rustic character tells us of 'the Savings Banks, the Loan Societies, and Benefit Societies as he robbed so shamefully, or the poor firesides he deprived of their hope and comfort by carrying off the little they'd scraped together by the labour of a long life' (iii. Ch. 7).[3]

It is to the private banks that the Victorian novelist turns when he wishes to weave the story of a banker into his fiction. The Victorian novelist was concerned with a 'plot', in which there were to be secrets,

[2] This kind of bank could, and did, get into difficulties by accepting bad business, making imprudent advances,. discounting doubtful paper, and occasionally re-discounting their own bills in order to increase their lending facilities. (See Clapham, *Economic History*, I, p. 515.) Comparatively few joint-stock banks were involved in fraud.

[3] Loan Societies were established to lend money at interest to the industrial classes and to receive payment by instalments. Their constitution and practice were regulated by the Loan Societies Act of 1840. Benefit (Friendly) Societies were voluntary associations for the mutual relief and maintenance of members in sickness, old age, or distress. The first Friendly Society Act was passed in 1793, and voluntary registration was offered in 1846. It will be realized that to involve such societies in ruin through fraud was particularly heartless and reprehensible, as they existed to husband the meagre resources of an underprivileged class of people, to whom ruin meant beggary.

known only to the author, which could be dramatically revealed when the time was ripe. Where financial chicanery was to figure prominently, there had to be no awkward shareholders demanding the figures, or incorrupt auditors periodically descending upon the premises. The swindler is seen to be alone, sole master of his own tormented conscience; if he makes a conspirator out of one of his employees, such a figure will only add to his burdens.

It will be appreciated that the various villains and monsters who adorn the fictional banks of our novelists were not typical of the class. In the vast majority of cases it was imprudence, not fraud, that led to crashes, and many bankers who failed in the crisis years were themselves victims of the imperfect financial system of their day. This is largely true of the majority of the country banks that failed in 1825 and 1826. The London private banks enjoyed a high reputation for probity and soundness, which was, in the main, justified. There were sixty-three such private banks in the capital in 1825, some of which had been established since the seventeenth century, though, as will be seen, long establishment was not always a guarantee of honesty.

A reassuring harmony existed between the London private banks and the Bank of England. The Bank was quick to guarantee their paper against its bullion reserve, and in time of crisis would come to the aid of these loosely associated houses. It kept their balances, and furnished them with gold or notes as required. Their demand for gold was generally small, sufficient to stock the tills, but together they were very considerable holders of Bank of England notes.

Continuing memories of the follies, frauds, and failures of 1825 were reflected in a number of Victorian novels in preference to later, equally memorable, commercial crises. Thus in Miss Mulock's *John Halifax, Gentleman* (1856), the affairs of Jessop and his bank are particularly accurate re-creations of the events of 1825, although, at the time the novel was written, the country had been scandalized and alarmed by a series of sensational banking frauds. Miss Mulock depicts Jessop as a hapless victim of circumstance, and Charles Kingsley had earlier adopted a similar attitude in his novel *Yeast*, which appeared in serial form during 1848.[4] In this work, Smith the banker is ruined as an indirect result of railway speculation:

[4] *Yeast* (1851) appeared first in serial form in *Fraser's Magazine*, from July to December 1848. Quotations in the text are from the Eversley Edition of the *Works* (1902). From the date of *Yeast*, it may be deduced that Kingsley's banker failed during the commercial crisis of 1847–8, which arose from the Railway Mania and food and money

Yes! The Bank had stopped. The ancient firm of Smith, Brown, Jones, Robinson, and Co., which had been for some years past expanding from a solid golden organism into a cobweb-tissue and huge balloon of threadbare paper, had at last worn through and collapsed, dropping its car and human contents miserably into the Thames mud. (Ch. 14, p. 265.)

These are highly effective images, pointing to Kingsley's understanding of the dangers attendant on the change from a specie-based to a credit-based economy. Old Smith is a prisoner of the system in which he operates: although a 'veteran Mammonite', he has a 'kindly and upright practical mind' (p. 210). Not all fictional bankers are kindly and upright, and in real life there were criminal and fraudulent bankers whose spectacular defalcations afforded vivid and long-remembered models for the creation by novelists of a whole gamut of financial villains. It is valuable to detail some of these cases, and to illustrate the nature of literary responses to them.

The twenties had seen the bursting of some 'bubble' banks, established merely as speculative enterprises, and these had wretchedly perished soon after the Bank had diminished its discounts. Some old-established houses, such as Sir Peter Poole, Thornton & Co., had fallen through imprudence and lack of caution, though no one would have regarded their principals as being in any way criminally culpable.

This decade did, however, witness two celebrated cases of downright fraud on a grand scale involving private banks, and there is little doubt that these were to influence the literary depiction of bankers for many years. The first was the fall, in 1824, of the firm of Marsh, Stracey, Fauntleroy & Co.; the second, four years later, was the criminal failure of Remington, Stephenson & Co.

The first of these two cases was the *cause célèbre* of banking history in the early decades of the nineteenth century. Marsh, Stracey, Fauntleroy & Co., bankers, of Berners Street, had been established in 1792 by a Mr Marsh and a Mr Sibbald,[5] neither of whom had any

panic of that period. Seventy-eight British and European banks failed during 1847–8. Only one London bank, Cockburn & Co., suspended payment; but fourteen English provincial banks failed: Lancelot's uncle's bank may be, as it were, a fictional fifteenth. The figures given are computed from the List of Failures appended to D. Morier Evans, *The Commercial Crisis 1847–1848* (1849), p. lxxxix.

[5] The account following is compiled from Morier Evans, *Facts, Failures and Frauds* (1859), and from the contemporary narration of the case in Andrew Knapp and William Baldwin's *New Newgate Calendar* (n.d.), in the edition produced by the Folio Society, London, 1960, pp. 114–22. In the statement made at his trial, Henry Fauntleroy

previous experience of banking, or could bring sufficient capital into the enterprise. They very quickly amassed debts of £20,000, probably through a combination of naïvety and bad management. In 1796, therefore, two experienced clerks, Mr Stracey and Mr Fauntleroy, were taken into the firm, and the two senior partners embarked upon an optimistic round of speculation.

In 1800 Mr Fauntleroy's son Henry, a clerk in the house, was appointed managing partner, and on the death of his father in 1807 he branched out into brick-making and building, offering generous accommodation to merchants and others engaged in these trades. Affairs did not improve, and many worthless bills were renewed. By 1810 the original loss had swollen to £170,000, and it was at this stage, and without his partners' knowledge, that Henry Fauntleroy began to appropriate customers' trust money and securities to support what was in reality an insolvent firm. All demands were promptly paid, and liberal advances offered, so that the house increased in commercial and public esteem.

In 1816 the building trade was severely hit by depression, and the failure of client undertakings absorbed a further £100,000 of customers' money.[6] At this stage Fauntleroy did a very curious thing. He wrote a secret memorandum,[7] in which he confessed that he had forged powers of attorney, and, without the knowledge of his partners, had sold out a total of £178,754 in securities. Having in some strange way relieved his conscience by this secret written confession, he continued his fraudulent practices for a further seven years, misappropriating investments, and applying the resultant sums to the general funds of

said that his father had been a founder of the firm (Knapp and Baldwin, p. 117); the statement in the text is based upon Morier Evans, *Facts, Failures and Frauds*, p. 107.

[6] There had been a typical commercial boom followed by collapse in 1816, when the building trade was particularly hard hit. It suffered again severely during the 1826 crisis. Mr Merdle, in *Little Dorrit*, was 'in everything good, from banking to building'; this could be an unfortunate combination of interests.

[7] 'In order to keep up the credit of our house, I have forged powers of attorney, and have therefore sold out all these sums (amounting, as per subjoined list, to £178,754.6*s*.4*d*.) without knowledge of any of my partners. I have respectively placed the dividends, as they became due, to account, but I never posted them. — H. Fauntleroy. 7th May, 1816.' (Morier Evans, p. 107.) The wording is rather different in the *New Newgate Calendar* version (p. 116), and the following words, not recorded by Morier Evans, are added: 'The Bank began first to refuse to discount our acceptances and to destroy the credit of our house: the Bank shall smart for it.' It is very probable, though, that Morier Evans had had sight of the original.

the bank. During this period, the reputation of Marsh, Stracey, Fauntleroy & Co. continued to be of the highest standing.

On 10 September 1824 Fauntleroy was arrested for fraudulent disposal of a batch of securities worth £10,000. A run on the bank took place immediately, resulting in suspension of payments and then bankruptcy. The house failed with liabilities of £799,028. At his trial, which took place on 30 October, Fauntleroy was specifically indicted for an earlier forgery of 1815, which was linked with the capital offence of uttering, as it was part of a series of transactions defrauding the Bank of England of £360,000. Fauntleroy, in view of his written confession of 1816, which had been found among his papers, was inevitably found guilty. On 20 November 1824 he was hanged at Newgate, watched by a largely sympathetic crowd numbering about one hundred thousand.

One celebrated literary reflection of Fauntleroy is worthy of examination, if only to show how an author's fertile imagination can stray so far from reality as to produce a character as unconvincing as it is fatuous. It is generally acknowledged that Bulwer's villainous banker, Richard Crauford, in *The Disowned* (1828), is a literary treatment of Fauntleroy.[8] Certainly Crauford appears to be a banker, though nowhere in the novel is this stated unequivocally, and the description of his business owes more to rhetoric than to commercial expertise: 'cool, bland, fawning, and weaving in his close and dark mind various speculations of guilt and craft, he sat among his bills and gold like the very gnome and personification of that Mammon of gain to which he was the most supple, though concealed adherent.'[9]

There are clearly no virtues here, and we are not surprised to learn that Crauford was 'fond of the toiling acquisition of money', and 'equally attached to the ostentatious pageantries of expense' (ii, p. 148). The charges of labouring in the service of Mammon, and being over-fond of the externals of extravagant living are frequently levelled by

[8] 'Lytton next produced a crop of stories in which topics from the *Newgate Calendar* are handled in the spirit of the "novel of sensibility". The highwayman with a heart of gold, the forger with regrets, the murderer with culture and a conscience . . . Lytton's offenders, though studied from actual Fauntleroys and Arams, carefully contrive to lose reality and forfeit sympathy. Their theatric habit, or rather the author's, is inveterate, and so is his trick of false extenuation, and of dressing up squalid motives in lurid or lofty language.' (Elton, ii, p. 189.)

[9] *The Disowned* (1828), 2nd edn., 3 vols., 1829; vol. ii, p. 148. Page references in the text are to this edition.

authors against their less savoury commercial figures, and Bulwer is not, of course, unique in making them here. What is interesting is that Crauford's ostentation seems to have been lacking in Fauntleroy. He declared at his trial that charges of personal extravagance were unfounded, and it is certain that his house at Lambeth was a modest home for a man of his station. He was particularly vehement in refuting these charges, which had been elaborated by the sensational press of the day, and it would seem that Bulwer was more interested in a man who uses the results of his fraud for his own personal gratification than, as was the case with Fauntleroy, to avert the closure of the banking-house.[10]

Bulwer is content to depict Crauford's moral character in terms of the popular press: 'His loves, coarse and low, fed their rank fires from an unmingled and gross depravity.' (II, p. 149.) In a statement made at his trial, Fauntleroy commented very strongly on the 'cruel and illiberal manner in which the public prints have untruly detailed a history of my life and conduct.'[11]

In the shadow of certain death, and with no possible way of justifying his financial crimes, he passionately defends his general 'respectability', in terms which are both compelling and pathetic:

> I have been accused of crimes I never even contemplated, and of acts of profligacy I never committed, and I appear at this bar with every prejudice against me, and almost prejudged![12]

> Gentlemen, I have frailties enough to account for. I have suffering enough, past, present, and in prospect; and if my life were all that was required of me, I might endure in silence, though I will not endure the odium on my memory, of having sinned to pamper delinquencies to which I never was addicted.[13]

A man who thus vehemently defended his honour when faced with the gallows, and one who had been moved to write a secret confession that ultimately hanged him, would surely have made an intelligent and complex study for the exercise of an author's genius. Remembering how Bulwer at least attempted such a study four years later in *Eugene Aram* (1832), it is disappointing to find Fauntleroy transmuted to the silly Crauford in the earlier novel.

[10] Knapp and Baldwin, pp. 118–19.
[11] Ibid., p. 116. [12] Ibid., p. 118. [13] Ibid., p. 120.

Crauford, we learn, is a master hypocrite, an ostensibly charitable churchgoer, who secretly gives vent to his contempt for 'your fine feelings, your nice honour, your precise religion—he! he! he!' (ii, p. 156.) We are not surprised when we learn that he plans to buy himself a peerage through bribery. His desire is ultimately granted, and his response—'He! he! he!'—is by now not entirely unexpected. This grotesque parody of Fauntleroy is as far removed from a real-life banker, honest or fraudulent, as Mr Copperas, the stockjobber in the same novel, is from the genuine frequenters of the Stock Exchange.

Equally removed from reality are the details of Crauford's supposed defalcations. He is described as having engaged in 'an extensive scheme of fraud . . . which for secrecy and boldness was almost unequalled' (ii, p. 154); but the reader is never given any details of this 'extensive scheme'. Bulwer does, however, insist that the 'scheme' is too great for one man to handle, and that it is imperative for Crauford to secure a confederate.[14] Crauford, however, already has such a confidant, the obsequious Bradley, described in the novel as 'the partner of his fraud' (iii, p. 262), and it is thus very obtuse of him to seek the co-operation of a particularly incorruptible character, Mordaunt, in his nebulous enterprise.

When Crauford is finally unmasked, the 'details' of his fraud are revealed in the following words:

> Inquiry developed, day after day, some new maze in the mighty and intricate machinery of his sublime dishonesty . . . houses of the most reputed wealth and profuse splendour, whose affairs Crauford had transacted, were discovered to have been for years utterly undermined and beggared, and only supported by the extraordinary genius of the individual, by whose extraordinary guilt . . . they were suddenly and irretrievably destroyed. (iii. pp. 326–7.)

It was possible for leading merchant houses to be destroyed in times of commercial crisis by the failure of imprudent bankers, but there was nothing 'mighty and intricate' about the 'machinery' involved. The most likely fraud that Crauford, a private banker, could have committed was the common one of misappropriation of private securities, not a 'mighty and intricate' matter to effect, as may be seen from the details of Fauntleroy's fraud, which were readily accessible to Bulwer as to us.

[14] In this detail the fictional banker departs radically from Fauntleroy. It is certain that the latter acted entirely alone, without the knowledge of his partners or counting-house staff.

It would have been most unusual for a large mercantile house to have based its many transactions on accommodation afforded by one private bank; nor would it have permitted its capital to lie fallow for more than a few months, certainly not long enough for Crauford to exercise to the full his 'extraordinary genius'. Bulwer here shows that he misunderstands the mercantile uses of capital in his own time.

Four years after the unmasking of Fauntleroy the firm of Remington, Stephenson & Co. failed, with liabilities of £508,696. The principal of this firm, Roland Stephenson, was MP for Leominster, and Treasurer of Bartholomew's Hospital. He lived in grand style at his country seat, Marshalls, in Essex, and was highly regarded in both City and county circles. But his prestigious way of life was supported entirely by fraudulent use of the bank's money, and when it seemed that detection was imminent Stephenson fled to Savannah, Georgia. He had robbed his customers of more than £200,000, and to assist him in his flight he took with him a further £50,000. The fall of this bank occasioned acute distress, particularly among the many widows and annuitants whose funds Stephenson managed.[15]

The case of Roland Stephenson furnishes a celebrated example of the reality underlying the many instances of bankers and brokers in nineteenth-century fiction who flee to America, as a secure haven for the enjoyment of their ill-gotten gains. It is quite likely that Peacock was recalling Stephenson's flight to Savannah when, in *Crotchet Castle* (1831), he created 'Mr. Touchandgo, the great banker'. One foggy morning, we learn, 'Mr. Touchandgo and the contents of his till were suddenly reported absent', and it soon transpires that he, too, has fled to the New World. In a letter to his daughter, Touchandgo explains the motives for his flight, and, allowing for Peacock's delightfully dry satire, one can discern an undercurrent of possible truth to life in his explanation. Roland Stephenson may well have reasoned along these lines: 'This was my position: I owed half a million of money; and I had a trifle in my pocket. It was clear that this trifle could never find its way to the right owner. The question was, whether I should keep it, and live like a gentleman; or hand it over to lawyers and commissioners of bankruptcy, and die like a dog on a dunghill . . . I decided the question in my own favour . . .' (Ch. 11).[16]

[15] Morier Evans, *Facts, Failures and Frauds*, p. 108.

[16] Two years after the appearance of *Crotchet Castle*, Harriet Martineau provided a similar profitable bolt-hole for her villainous fraudulent banker, Mr Cavendish, in her novel *Berkeley the Banker* (1833), No. 14 in the series *Illustrations of Political Economy*. Among

Men of Stephenson's stamp were able to reap the benefits of their criminal actions. Through the exercise of sufficient shrewdness and alertness, they could not only escape the very severe and rigorous laws under which they would have been judged, but also find in America a fresh field for the exercise of their talents. The absence of extradition agreements allowed these criminal fugitives to remain in affluent liberty, and their situation, contrasted with the despair and destitution of their victims at home, provided the novelist with effective material for cynical observations on an outrageous reality.

The dreadful fate of Henry Fauntleroy stands in a different category. Society was ready to see the death penalty for forgery removed, and his execution aroused considerable public revulsion. Acts of 1826 and 1837 abolished the death penalty for this and similar offences, but there can be little doubt that the public hanging of this banker kept the details of his crime more firmly in the public memory than would have been the case had he suffered imprisonment or transportation.

A second, even more disturbing series of criminal bank failures was to occur in the fifties, keeping this branch of commerce firmly in the novelist's purview. The Railway Mania of the late forties produced a dangerous *malaise* in commercial life, arising from a clash between consciousness of the rectitude of traditional standards of commercial morality and a desire to repeat the heady if dangerous successes of the railroad-share epoch. Very-old-established firms of high reputation began to engage in reckless speculation. Protected from rumours of 'unsoundness' by honourable trading-names, they heavily compromised themselves, and, perhaps inevitably, resorted to fraud.

This new and disturbing phenomenon, to which Morier Evans gave the very effective name 'High Art Crime', consisted of a 'series of financial and commercial delinquencies, in which persons of supposed elevated character were involved, that all the received tests of respectability seemed to be of no avail, and people literally could not tell whom they might trust'.[17] In particular, the years 1855 and 1856 witnessed several spectacular criminal failures in the banking world, and there can be no doubt that these events perpetuated the by then traditional depiction of bankers in the novels of the period. The former

other nefarious activities, Cavendish ran a forger's den, and when it seemed certain that justice must catch up with him, he emulated Roland Stephenson, 'in time to step quietly on board an American packet' (ii, Ch. 7).

[17] Morier Evans, *Facts, Failures and Frauds*, pp. 3, 4.

year saw the fall of John Sadleir's Tipperary Bank, described in a different context later in this book, and the criminal bankruptcy of Strahan, Paul & Bates, one of the most respected private banking-houses in the City. The year 1856 witnessed the closure of the Royal British Bank, and the prosecution of its directors. This was a joint-stock bank, established in 1850, and grossly mismanaged from the beginning, with a cavalier approach to the use of clients' money. Its ruined carcass was tugged between the Courts of Bankruptcy and Chancery, and within a year its wretched existence had been forgotten.[18]

It was, however, the fall of Strahan, Paul & Bates that was to have the most profound effect upon the public. They were to hear of a new and frightening commercial profligacy involving honoured and stable household names, and it is possible that the renewed distrust of bankers of whatever type engendered by this particular closure lasted until the end of the century.

Strahan, Paul & Bates was one of the oldest banking houses in London, having commenced business during the Commonwealth. At the time of its fall it was administered by three active partners: William Strahan, a lineal descendant of the founder; Sir John Dean Paul, Bt.; and Robert Bates, a salaried partner since 1841, and a pliable tool of the senior partners in their criminal activities. The two latter were true City aristocracy. Strahan, a man of sober and responsible private life, was highly esteemed in the City: 'to breathe a word of suspicion against his honesty would have been thought as unreasonable as to dispute the credit of the Bank of England.'[19] Sir John Dean Paul, son of the first baronet of the same name, was born in 1802, and educated at Westminster School and Eton. Of aristocratic background and high standing as a banker, he was also accounted a very religious man: in 1846 he had published *Harmonies of Scripture*.

The novelist Mrs Catherine Gore was a ward of the elder Sir John Dean Paul, and, not unnaturally, she banked with the firm. In 1843 she dedicated her novel *The Banker's Wife* to her guardian. She was anxious that old Sir John should not feel hurt by her depicting a rascally private banker, who could possibly be taken as representative of a class. Her protagonist, Richard Hamlyn, is seen fraudulently converting clients' securities, and one may readily assume that she had Fauntleroy

[18] This was an exceptional case of fraud executed within the stringencies of joint-stock banking rules. See the complex account in Morier Evans, *Facts, Failures and Frauds* (Ch. 7).

[19] Ibid., pp. 109–10.

in mind when she created this character. Consequently, in the Dedication to the first volume of her novel, she writes:

Dear Sir,
I cannot more strongly mark that the following pages are intended to exhibit the failings of an individual, not as an attack upon a class, than by placing at the head of my work the name of one who, ancestrally connected, for the last two centuries with the banking profession in a house of business which has existed in the same spot since the year 1650, has added to its distinctions in his own person . . .

It is an impressive pedigree, masking a corrupt and rotten sham. Even while Mrs Gore was lauding the stability of her banker, the house was involved in hazardous, ultimately ruinous, enterprises, and had been guilty of malpractices since the early years of the century. By 1843, the date of this Dedication, the house was virtually insolvent.

Dean Paul senior died in 1852, and was succeeded in the baronetcy by his son of the same name. It was this second Sir John Dean Paul who was to appear on trial for fraudulent conversion at the Central Criminal Court in October 1855, having failed to seek refuge in bankruptcy.[20] The fall of this house adds a special irony to Mrs Gore's dedication. Her deposits in the bank of her trusted guardian were irretrievably lost, and this prolific and hard-working novelist emerged the poorer by the enormous sum of £20,000.[21]

The crash was occasioned by the firm's making frequent calls upon discount houses,[22] an unusual and imprudent action which led to a

[20] By openly confessing their misuse of clients' money during a voluntary examination in bankruptcy, the partners hoped to avoid subsequent prosecution for a felony, under certain terms of the Fraudulent Trustees Act (7 and 8 Geo. IV). This subterfuge was disallowed, and the trial proceeded (*Facts, Failures and Frauds*, p. 125 f.).

[21] *DNB* (viii, 1922 edn.), evidently confuses the two Sir John Dean Pauls. 'It is a curious fact that in this work [i.e. *The Banker's Wife*] there is described such a dishonest banker as Paul himself afterwards proved to be. By the bankruptcy of Strahan, Paul, and Bates, on 11 June 1855, Mrs Gore lost £20,000.' (See volume cited, under *Gore, Mrs. Catherine.*) The wording of the entry certainly suggests that the bankruptcy involved the Paul mentioned in the first sentence. The entry for Mrs Gore in Kunitz and Haycraft, *British Authors of the Nineteenth Century* (New York, 1936), derives from *DNB*, and adds further to the confusion: '. . . it depicted an absconding banker, and apparently stimulated her guardian to go and do likewise—soon after, he absconded with £20,000 of her fortune!' (p. 255). Her guardian had died in 1852, and his son did not 'abscond': he surrendered himself at Bow Street on 21 June 1855. (Richard Hamlyn, the 'absconding banker' in the novel, did not 'abscond' either.) See Morier Evans, *Facts, Failures and Frauds*, Ch. 4, and his *History of the Commercial Crisis, 1857–1858* (1859), pp. lxxviii ff.

[22] Including Overend, Gurney & Co., 'the banker's bankers'. This great discount house itself failed during the crisis year of 1866. (See Clapham, *Economic History*, ii, pp. 374 f.)

run on the bank on 8 June 1855. The bank was listed in the *Gazette* on 11 June, having admitted liabilities of some £750,000. One client, Dr Griffiths, Prebendary of Rochester, lost Danish securities valued at £22,000; he was, however, rich enough to take out warrants for the partners' arrest for felonious disposal. At the Central Criminal Court on 26 October 1855, all three partners were found guilty of a felonious conversion under the terms of the Fraudulent Trustees Act, and they were each sentenced to fourteen years' transportation. Ironically, the shattered business was taken over by the London and Westminster Joint-Stock Bank, which opened a branch office in their Strand premises.

The misuse of Dr Griffiths's securities was merely the tip of the iceberg upon which the partners foundered. As early as 1816, the then partners had been personally indebted to the bank for sums totalling £29,000; ten years later, two of them owed between them £53,000. The only security offered for these vast sums was a joint note. This was not only morally reprehensible but also very bad and perilous business, as much of the money came from the deposits of clients.[23] When the second Sir John Dean Paul inherited the senior partnership in 1852, the bank was already hopelessly insolvent, with a deficiency of some £72,000. The partners then launched into speculative enterprises that were to end in hopeless ruin. Investment in a mining venture swallowed a further £45,000. Massive advances were made to the civil engineering firm of Gandell & Co., for railway and drainage schemes in France and Italy, all of which failed. In desperation, Strahan surrendered his private fortune of £100,000 to the bank, but this too was swallowed up. When the house finally crashed, Gandell & Co. still owed it £289,000, and so fell with them. This, then, was the desperate secret background to the criminal acts of felonious conversion that brought the partners as convicts to the hulks.

The fall of these men was calamitous in a way not easily imagined in present-day society. These men were gentlemen, highly regarded, and with a valued City pedigree. Now suddenly they were felons, and could have shared a corner of the hulks with

[23] 'The connection, though not extensive, was one that bankers generally consider the best paying. It consisted chiefly of members of the aristocracy, and wealthy commoners, who habitually keep large balances in their banker's hands.' (Morier Evans, *Facts, Failures and Frauds*, p. 111.)

Abel Magwitch.[24] Any compassion for them was quenched by remembrance of the plight of their victims. Morier Evans regarded their sentence as 'a wholesome example to deter others', and the following words by the same writer, describing Dean Paul's reaction to the sentence, speak eloquently of prevailing attitudes:

> his eye wandered among the crowd, until it fell upon the countenance of the prosecutor [sc. Dr Griffiths], and having attracted his attention, he gazed upon him with a look which, if the reverend doctor had not been upheld by the self-consciousness of imperative duty, might have raised regretful feelings in his mind. (Op. cit., Ch. 6, p. 123 f.)

The fall of Strahan, Paul & Bates was a *cause célèbre* of nineteenth-century commercial history, and references to it are found beyond the narrow bounds of the business community. After the partners had been sentenced, doggerel verses became popular, particularly a doleful elegy beginning:

> Paul, Strachan [*sic*] and Bate,
> Hard is their fate,

and a rougher song, supposedly composed by a convict, which commenced with the lines:

> If I'd been a partner in a bank
> I shouldn't have been working at this here crank.[25]

Writing in 1862, Henry Mayhew described a class of begging-letter writer that claimed to have been ruined by 'the suspension of Haul, Strong and Chates'.[26] The decorously altered names are readily identifiable, and show how the affair was still firmly in the public mind at that time. Ten years later, in his *London, a Pilgrimage* (1872), Blanchard Jerrold, commenting on the ever-yawning precipice between rags and riches, spoke of

[24] Of Richard Hamlyn Mrs Gore writes: '. . . this man of universal credit was but a more polished, more cautious, more solid swindler, in the amount of thousands, where swindlers in the amount of tens or hundreds are sentenced to the hulks.' (*The Banker's Wife* iii, Ch. 3.) These words are disconcertingly like a prophecy.

[25] The verses are recollected by two correspondents in *Notes and Queries*, 7th Ser., Vol. X (1890), the first in the issue for 27 Sept. (p. 247), and the second in that of 18 Oct. (p. 312). It is interesting to note that the editor, in this latter issue, declares that 'Other replies are acknowledged, but further discussion is not invited'. This may have been out of deference to, or at the request of, Sir John's son and heir, Sir Aubrey John Paul, Bt., who had succeeded his father in 1868.

[26] P. Quennell (ed.), *London's Underworld* (1950), p. 349.

the brave men who have handled an office broom in the beginning, and ended the possessors of enormous wealth, and the objects of general respect. In the list opposite . . . are the names of men who began with wealth and ended in disgrace and rags . . . the Sir John Dean Pauls, the Redpaths, and the Roupells. (p. 138.)[27]

Sir John Dean Paul died at St Albans on 7 December 1868, aged 66. *The Times* afforded him an obituary of a single paragraph, in which it referred to him as 'a once notorious character' and alluded briefly to his crime and its punishment.[28]

When Dean Paul and his associates were not engaged in defrauding their clients, they followed assiduously the business of private banking, as it was practised in the early decades of the century. Private bankers received deposits for investment, and, in the case of the country banks, provided in addition a convenient circulating credit through the issue of their banknotes. They also acted as trustees of entailed estates and as executors. Mrs Gore, in *The Banker's Wife*, gives an accurate characterization of the well-established private houses of her day: 'The firm of Hamlyn & Co., if unsupported by enormous capital in the private property of the partners . . . was trebly secure in its own moderation, steadiness and good renown.' (ii, Ch. 2, pp. 69–70.)

It was in some ways easier for the London private banks to parade these virtues than it was for their country counterparts. Country banks, being banks of issue, were always prone to printing paper money without having the gold necessary to redeem it. The Bank of England, despite the fiction printed on its notes, was not bound to redeem its notes for gold, and in the years prior to an earlier crisis in 1816, had issued so many notes that people doubted whether they would ever be able to honour them. The country banks were not slow to emulate their big sister, but they had not, of course, her resources, so that a certain hand-to-mouth precariousness developed, well described by Harriet Martineau in *Berkeley the Banker* (1833): 'the more paper money the Bank of England issued, the more were the proprietors of other banks tempted to put out as many notes as they dared . . . and some cheats and swindlers set up banks, knowing that they should never be able to pay . . .' (i, Ch. 3.)

[27] It should be remembered that the expression 'disgrace and rags' is to be taken literally: the reality in part explains the frequent suicides that followed the detection of fraud committed by financial magnates of the period.

[28] *The Times*, 15 September, 1868, p. 4.

One of these cheats and swindlers is her own fictional Cavendish, a prosperous corn, coal, and timber merchant, who sets up a country bank with the sole object of defrauding his customers. Relying on his till-capital to cover his liabilities, he retains a precarious confidence by wide issue of his notes, 'without being particularly scrupulous as to whether he should be able to answer the demands they might bring upon him' (i. Ch. 4). Adding high-class forgery to his other dubious talents, he disappears when the time is ripe, and becomes a luminary of the New World.

Richard Hardie in Charles Read's *Hard Cash* (1863) is another country banker who operates in the same precarious manner. He had rescued the bank from insolvency during the 1825 crisis, an act which gave him high public standing. This he uses as a cloak for a whole series of dishonest acts, fantastically conceived in true Readean fashion, but hardly incredible to readers with memories of High Art Crime.

The examples of fraud and desperate mismanagement given so far from fact and fiction indicate how absolutely essential it was for the Victorian banker to seem above reproach, impeccable in his public and private life, and aloof from any breath of scandal. Countless hundreds of such men possessed such honour and integrity, and practised their vocation with a clear mind; the dishonest and the fraudulent, however, in order to appear indistinguishable from their honest colleagues, adopted a mask of calm probity that hid quite appalling desperation and mental anguish.

The hallmarks of a banker's public character were an outwardly grave and calm aspect,[29] and a shunning of personal ostentation. While maintaining a standard of living consonant with his status as a gentleman, he was careful not to allow himself to be tainted with the flashy vulgarity of the 'new men' of the age. In a highly hierarchic society, the maintenance of high social standing was essential to a sound public character. Mrs Gore provides a clear statement of the accepted mores of a London banker of her day:

A London banker, having a handsome stablishment in town, is held bound to re-assemble his domesticities about him, as soon as may be after the meeting of parliament. It would 'look odd' were his wife to be without an opera-box

[29] Mrs Gore describes her banker, Hamlyn, as 'a man eminently qualified to figure to advantage on a tombstone; he had never been suspected of a vice, or accused of a failing . . .' (*The Banker's Wife* i, Ch. 2. p. 34). Charles Reade's Richard Hardie was 'Dignity in person', with a 'grave, calm, passionless voice' (*Hard Cash*, Ch. 16).

during the season . . . It would 'look odd', were his pew in Mary-le-bone Church to be empty . . . and a banker is bound to eschew all and any thing that 'looks odd'. Everything about him, both in public and private life, should be as even as the balance of his books.[30]

This passage shows how such men of the upper middle classes were expected to observe the London 'season' in all its ritual detail. They would maintain a London house for that purpose, as well as an 'estate' in the country. Sir John Dean Paul resided at Nutfield, near Reigate; Roland Stephenson lived luxuriously at Marshalls, his Essex seat, supported there by £200,000 of his customers' money. The fictional Richard Hamlyn keeps a sumptuous town house in Cavendish Square, and an estate, Dean Park, in the Home Counties.

The fine line separating tasteful affluence from vulgar ostentation gave rise to a certain ambivalence in the writings of the period towards the external signs of wealth exhibited by these magnates. Speaking of Strahan and Dean Paul, Morier Evans had commended the lack of 'wasteful or wanton extravagance' in their private life, and applauded their 'liberality of expenditure, becoming the station of society in which they moved'.[31] It was surely difficult to decide where 'liberality of expenditure' ended, and 'wasteful or wanton extravagance' began.

In fiction the 'elegances of life' were almost invariably presented as a vehicle for moral condemnation, and as a fairly obvious symbol of the flashy obtrusiveness of the parvenu financier upon the fabric of English life. Mrs Gore, for instance, introduced a silver wine-cistern into Richard Hamlyn's household effects, which is mentioned on more than one occasion to emphasize the opulent but somehow vulgar quality of his domestic life; and on the night before he fights a fatal duel, Hamlyn's eye roves fondly over the 'gaudy objects' for which he had 'perilled the credit of an honest name'.[32] Bulwer's Crauford had done likewise, casting a 'gleam of gladness' upon 'four vases of massy gold' shortly before his arrest and execution.[33] The glittering drawing-rooms and dining-rooms, so carefully described in this type of novel as sumptuous or gaudy according to one's current prejudice,

[30] *The Banker's Wife* ii, Ch. 6, p. 203.
[31] Morier Evans, *Facts, Failures and Frauds*, pp. 108–9.
[32] *The Banker's Wife* ii, Ch. 6, p. 213; iii, Ch. 4, pp. 159–60; iii, Ch. 6, p. 237.
[33] *The Disowned* iii, Ch. 25.

were the setting for those dinners and receptions at which Society patronized the man of business, while secretly despising him.[34]

The ambivalence towards external signs of wealth applied also to the banker's place of business. The public expected a staid, old-fashioned counting-house, although a modest private residence would have set the rumours flying. Thus the premises of Richard Hamlyn afford a telling contrast to his opulent house in Cavendish Square.

> The house was of dingy brick, with low-browed, smoke-stained ceilings, and desks and counters of coloured mahogany; unlike those gorgeous banking-houses of the day (resembling gin-palaces in more particulars than one) which seem to have thriven, like parasite plants, out of the substance of others.[35]

The 'gin-palaces', the 'gorgeous banking-houses', were an amalgam in Mrs Gore's mind of the gaudy shams of the 1825 era, and, particularly, of the imposing headquarters of the new joint-stock enterprises. Dinginess seems to have been a guarantee of virtue, though hindsight shows that many of the 'gin-palaces' would have provided far greater security for savings than the staid frontage of Strahan, Paul & Bates, or the gloomy repose of Hamlyn's.[36]

The banker had to be constantly watchful that his reputation for sobriety in his profession and opulence in his private life should not at any time be called in question. Hamlyn is obsessively aware of this need. When his wife suggests domestic economies, he exclaims: 'Are you mad? . . . The men are in the dining-room removing the breakfast things. If Ramsey should hear you—'. When his wife later suggests that it would be wise to let their London house for the season, he retorts 'with a bitter smile' that this would be the best way to rouse the

[34] The flashiness complained of seems to have been more than metaphorical. R. H. Warnum, in his essay 'The Exhibition as a Lesson in Taste' (1851), castigates the prevailing English taste for 'boiled out' silver for table decoration. '. . . when the dead white thus produced is combined only with burnished portions, the sole effect of a work is a mere play of light without even the contrast of shadow. The result is a dazzling whiteness; pure flashiness in fact . . .'. A large silver centre-piece by Hunt and Roskell is said to display 'the characteristic flashiness of the prevailing home taste of the present day' (*The Art Journal Catalogue of the Great Exhibition, 1851*, p. I*** f.).

[35] *The Banker's Wife* ii, Ch. 2, pp. 67–9.

[36] Dickens uses the same dinginess to describe Tellson's Bank in *A Tale of Two Cities* (1859): '. . . the triumphant perfection of inconvenience . . . a miserable little shop, with two little counters' (ii, Ch. 1). The premises of Strahan & Co. in the Strand were unexceptional, and Dickens's own bankers, Coutts & Co., established at number 59 in the same street, were satisfied with a dignified and sober eighteenth-century frontage. (See M. V. Stokes: 'Charles Dickens: a Customer of Coutts & Co.', in *The Dickensian*, Jan. 1972.)

anxieties of his customers, and 'to place the names of Richard and Bernard Hamlyn in the Gazette'.[37]

Such considerations constantly exercised the minds of countless men of business who strove to remain solvent, since any sign of obvious retrenchment could bring instant ruin. This need was a severe strain for the honest man facing bankruptcy; for the fraud, it was intolerable. Hamlyn upbraids his wife for making allusions to 'what cannot be even whispered in the stillest watches of the night'. Of Richard Hardie's lonely struggle to remain solvent, Reade tells us that 'the price of one grain of sympathy would have been "Destruction".'[38]

Just as the nineteenth-century novelist was sharply aware of the external public image of the bankers and financiers when he chose to depict them in his novels, so too he was able to communicate very effectively the internal, private thoughts and feelings of such figures. They are inevitably seen as men totally dedicated to their profession, so that even the honest men become devotees of Mammon, their lives moulded by the pursuit of Gold.

The public results of this private dedication to Mammon are a principal theme in novels such as *The Banker's Wife*; but of equal importance to the novelists is its effect upon the family life of the banker. There is a strong depiction of a perversion of normal family values, and a growing intellectual alienation between the banker, and his wife and children. Mrs Hamlyn and her family have been blighted by Hamlyn's transmutation from a carefree young man to a sombre slave of Money.

> His wife knew him to be averse to all display of sensibility; his children were early taught that he detested noise; and the banker's house was, consequently, characterised by the silence, coldness and dullness of the Great Pyramid.[39]

This is the same literary response that one sees in Dickens's portrayal of the spiritless and crushed first Mrs Dombey, and in the pallid, acquiescent Mrs Gradgrind.

An important depiction of alienation in these novels is the creation of a younger generation who turn aside from 'money-getting' to other, more noble, pursuits. Hamlyn's son Henry, turning his back on what he calls 'vulgar strife and chicanery', becomes a scholar at Cambridge. Similarly, Richard Hardie's son in Charles Reade's *Hard Cash* (1863), develops into a first-rate scholar and athlete at Oxford, where he is

[37] *The Banker's Wife* ii, Ch. 6, pp. 204, 209.
[38] *The Banker's Wife*, ii, Ch. 6, p. 210; *Hard Cash*, Ch. 8.
[39] *The Banker's Wife* i, Ch. 2, p. 35.

able to fulfil himself as an individual, free from his father's engrossment in banking enterprises.

Charles Lever gives a wider depiction of a family turning away from a financial ambience in *The Bramleighs of Bishop's Folly* (1868). Colonel Bramleigh is a sound and respectable man married to a lady of birth. Although loyal to her husband, and contemptuous of the various aristocratic toadies who pretend to cultivate him, she none the less lives, with his sanction, in Italy, an ornament of the English colony in Rome. Lever thus cleverly depicts the incompatibility, not of two people, but of two social systems.

It is the Bramleigh children who illustrate the problems attending a rich bourgeois family in achieving a personal identity. They are constantly aware of the structures of society. Temple Bramleigh reveres the aristocracy, and enters the diplomatic service, so that he may be constantly in the company of 'the best people':

> By the best people, I mean the first in rank and station. I am not speaking of their moral excellence, but of their social superiority, and of that pre-eminence which comes of indisputable position, high name, fortune, and the world's regards. These I call the best people to live with. (Ch. 3.)

Jack Bramleigh's view of life is more radically democratic: 'I want to associate with my equals. I want to mix with men who cannot overbear me by any accident of their wealth or title.' (Ch. 3.) Only Ellen and Augustus seem content with their lot, accepting their placing in the middle classes of society. Ellen declares: 'we are great fools in not enjoying a very pleasant lot in life instead of addressing ourselves to ambitions far and away beyond us.' (Ch. 4.)

It is surely safe to assume that the novelist's depiction of children striving after academic or genteel and sheltered lives reflects a true contemporary situation. The possession of great wealth without titled rank must have produced such crises of identity, and a desire to gloss over the commercial enterprise upon which that wealth had been founded.

The Victorians were overtly emotional people, more ready to give vent to anguish in words and tears than are their descendants. Men of business who fell into disgrace and ruin were frequently smitten with a fear and remorse utterly harrowing to the twentieth-century reader.[40] This emotional trait was much used by novelists to

[40] See in particular the account of John Sadleir, M.P., in Ch. 6 of the present work. It is still distressing to read Sadleir's confession, and the general effect of his words tallies exactly with the tone of the many soliloquies delivered by 'villains' in Victorian novels in their moments of personal crisis.

emphasize the inner torments of fraudulent business men. It could be very effectively developed by involving the protagonist with a cringing, fawning subordinate, who either knew or guessed his guilty secret, and who thus humiliated his master by a kind of sneaking and contemptible insolence.

Bulwer's Crauford had had such a confidant, Bradley; and Uriah Heep is but the most celebrated of these monsters, who are bound inextricably to a weak or guilty master. The defalcations of Mrs Gore's Hamlyn had not gone unnoticed by his head clerk, Spilsby, who had only to 'fix his eye searchingly and insolently on his master' to make the banker change countenance and speak incoherently. By the fourth chapter of Book Three of *The Banker's Wife*, Hamlyn is possessed by monomania: Spilsby is no longer a mere man but a monstrous creation of the banker's own guilty fear, a 'domestic serpent', whose 'pestiferous breath' induces 'a decay by slow poison . . . a gradual decay of the flesh and the spirit'.[41] It is professionally imperative for Hamlyn to conquer his shame and horror and to maintain his bland exterior pose, and this he is able to do until sudden death solves his problems. The tension created between public assurance and private turmoil is one of the most dramatically effective qualities of this novel.

Charles Reade's banker, Richard Hardie, is also made to ensure a feared and unwanted confidant, Skinner, who, like Spilsby, functions to dramatize his master's anguish. Corrupted by Hardie's amorality, he works to put his master in his power, and when the time is ripe informs Hardie that he knows all about his frauds. 'Each word of Skinner's was a barbed icicle to him . . . Hardie turned his head away; and in that moment of humiliation and abject fear, drank all the bitterness of moral death.'[42] Hardie, though, is a survivor, and Reade sees fit to remove Skinner by having him conveniently choked to death by fumes from his stove. Hardie, his mind unhinged, takes to begging in the street. When he dies, he is found to be worth £60,000. Considering the villainous nature of his career, he goes out in triumph.

A further subject that afforded good dramatic material for the novelist was the excitement following the news of a bank's closure, together with the physical run on the house by its panicking clients. It was shown earlier in the present work how Miss Mulock, always economical in her effects, brought a convincing realism to her account

[41] *The Banker's Wife*, iii, Ch. 4, pp. 134–5. [42] *Hard Cash*, Ch. 16.

of the run on Jessop's bank in *John Halifax, Gentleman* (1856). Mrs Gore's treatment of a rather different kind of closure is equally memorable. In the closing chapters of *The Banker's Wife*, Hamlyn's clerk, Spilsby, and Harrington, a relative of Hamlyn, with the noise of the frantic crowd ringing in their ears, decide to post up notices of closure. Harrington's blunt and forthright declaration of the true state of things demolishes at a stroke the façade of probity and dignified aloofness that had so completely masked the man of straw.

I see no use in attempting to keep up the farce! . . . To open the house for the despatch of business is wholly out of the question. . . . The fact is, that the firm was involved at the old man's death.[43] Ever since, instead of retrieving himself by self-denial and economy, Richard Hamlyn has been plunging deeper and deeper into the mire . . .

Reade, in his depiction of the run on Hardie's bank, concentrates into one page the initial and subsequent actions of the crowd, conveying in long, compound sentences the personal concern for suffering that is typical of this author. Free from overt moralizing, his itemization of specific forms of hardship is all the more compelling.

The climax of this page-long episode is a physical attack upon the bank and the adjacent house. For this, Reade conveys a deliberately detached lack of involvement in Hardie's plight, which is very clever and superior to any stylistic effect that Mrs Gore could have achieved.

A heavy stone was flung at the banker's shutters . . . it was the signal for a shower; and presently tink, tink, went the windows of the house, and in came the stones, starring the mirrors, upsetting the chairs, denting the papered walls, chipping the mantelpieces, shivering the bell glasses and statuettes, and strewing the room with dirty pebbles, and painted fragments, and glittering ruin. (*Hard Cash*, Ch. 24.)

It takes only the 'tink, tink' of a few pebbles to bring the gaudy house of cards crashing down. Reade lists those flashy objects which were the symbols of wealth and status[44] in order to demolish with them the rotten fabric that had sustained them.

[43] *The Banker's Wife*, iii, Ch. 7, p. 284. The 'old man' is Hamlyn's father. For the second time in this novel Mrs Gore seems to speak with a prophet's voice, as Strahan & Co. were certainly 'involved' at the time of the elder Dean Paul's death. One wonders how the then Mr Dean Paul reacted to reading these words in 1843.

[44] Mrs Gore had referred specifically to similar objects in her description of Hamlyn's feelings on the night before his duel (iii, Ch. 6). In *The Bramleighs of Bishop's Folly*, the character Augustus has this to say of such material possessions: '. . . luxury

None of these novelists ever forgets the plight of the victims of bank closures, particularly the poor, humble, and helpless. Mrs Gore, throughout *The Banker's Wife*, creates a number of situations that portray domestic felicity and contentment, but which at the same time carry chilly premonitions for the reader of the disasters to come. These people are not speculators: they simply entrust their incomes to Hamlyn, often on personal recommendation. Some are gentry, or military men, or parochial clergy; others are of humbler station. There are the children of a physician, an aged widow, and Miss Creswell, a governess, who invests her £1,000 with Hamlyn to secure a life annuity. These latter will now be given the choice between beggary and the workhouse.[45]

Reade's treatment is generally similar to that of Mrs Gore. Two characters in *Hard Cash* are early designed as victims of Hardie's closure; Captain Dodd, owner of the 'hard cash', and James Maxley, a market gardener. Dodd very foolishly deposits £14,000 with Hardie on private receipt, and one feels that some of the inconvenience that follows is his own fault. In any case, he ultimately retrieves his fortune intact. Maxley's is a sadder predicament. He banks with Hardie because he mistrusts 'Lunnon banks', but his is a foolish mistake. Hardie's, a country bank, is a bank of issue, so that Skinner is able to pay the illiterate Maxley in Hardie's own notes when the run commences, instead of in those of the Bank of England. When his wife sees Hardie's worthless notes, she suffers a heart attack and dies. Maxley loses his reason, and is confined to a madhouse.

In Chapter 24 of his novel Reade produces a series of deliberately detailed vignettes of hitherto unmentioned victims of the closure, in order further to arouse our indignation against Hardie. These portraits

and splendour . . . were the daily accidents of life—they entered into our ways and habits, and made part of our very natures; giving them up was like giving up ourselves, surrendering an actual identity.' (Ch. 33.)

[45] In the event Mrs Gore chooses to bring some relief to all the named victims of Hamlyn's fall. His son inherits a fortune connected with a 'Bombay Company', and Hamlyn's 'South American speculations' appear to have been successful (iii, Ch. 9). This money is applied to making appreciable amends to the creditors. Such unexpected good fortune belongs to fiction; Strahan & Co. ultimately paid its creditors 3*s*.2*d*. in the pound. (See J. F. R. Barker: 'Paul, Sir John Dean', *DNB* xv, 1922 edn.) Mrs Gore stresses one very dangerous result of the fall of a well-established bank, the lack of public confidence in banking as an institution. 'If *they* were insecure, who was solid? . . . More than one banking-house of the highest reputation had cause to rue the discoveries of that day!' (iii, Ch. 7, p. 308).

provide a cross-section of the various classes who used banks; they are the kinds of customer who were ruined by country closures in particular. Recalling how closely Reade researched his fiction, it is quite possible that some of these victims are based on actual cases.[46] We read of Mr Esgar, a 'respectable merchant', who is forced to decamp to America, thus passing on the effects of his own ruin to others. Four fishermen, the heroes of a sea-rescue, buy a new boat on the security of a public subscription lodged at Hardie's bank. With the money lost, they are unable to pay for the boat, and are consigned to prison. Dr Phillips, aged seventy-four, has sold his practice and retired, having placed all his money at the bank. He and his wife and daughter become paupers, and he ekes out a miserable existence in a neighbouring village. 'And so Infirmity crept about, begging leave to cure Disease.'

The novels examined in this chapter may be taken as typical products of an age that had witnessed many explosive events in the world of banking. The failures of 1825, and the sensational felonies of Fauntleroy and Stephenson in the same decade, taken in conjunction with the 'high art' criminal activities of Dean Paul and others in the fifties, ensured the novelist a plentiful supply of what Charles Reade once described as 'fertile situations', which for him were 'the true cream of fiction'.[47] The modern reader does well to remember that not all Victorian bankers were practitioners of 'High Art Crime', and that the English banking system enjoyed an international reputation for stability and integrity. Fiction, though, is not concerned with the commonplace, and there is no doubt that in general terms the 'fertile situations' furnished by criminal and other failures were legitimately handled by the novelists, and echoed those failures with considerable fidelity.

As we have seen, the basically honourable profession of banking could not prevent its occasional penetration by monumental rogues. It was, perhaps, inevitable that the business of Life Assurance, a comparative late-comer to the English commercial scene, should be similarly vitiated.

[46] '. . . his tales were founded on producible documents drawn from real life. He gathered, he pigeon-holed, he bequeathed for public inspection, great files of cuttings, references and other evidence . . . above all examples of cruelty or legal oppression, are the material that he draws . . . from his clippings.' (O. Elton, *A Survey of English Literature 1830–1880* (1927), vol. ii, pp. 225–6.)

[47] The *Bookman* xviii, p. 252: '. . . fertile situations are the true cream of fiction . . . with these supplied, any professional writer can do the rest'.

This enterprise may be said to have commenced with the opening, in 1765, of the Equitable Assurance Society, the first company to conduct Life business on lines familiar to us today. After a lull of some decades, there was a rapid expansion in the number of offices, 150 coming into being between 1800 and 1843. Before the passing of the Life Assurance Companies Act in 1870, control was rudimentary, and in the midst of the many excellent undertakings there were many rogues out to make a quick fortune from a business that made the realization of this ambition a realistic proposition.

4

Insurance Promoters: Dickens, Thackeray, and the West Middlesex Fraud

Nineteenth-century people were alert to the 'shady' nature of many Life and Annuity offices, as there had been a spate of criminal companies in the previous century. These had offered to pay annuities, with no lump sum, at some date in the future, so that their sole function had been to collect premiums. Touts were offered the equivalent of a year's premium to enrol new victims. Equally disreputable had been gambling insurances, whereby a man could insure the life of a public figure at, say 7 per cent, collecting if he survived. These policies were prohibited by an Act of 1774, and further legislation in 1777 made it obligatory to register in Chancery all documents relating to 'the pernicious habit of raising money by the sale of annuities'; however, the imperfect state of Company Law before the 1840s rendered any Government action largely ineffective.

Because of this lack of full legal control over joint-stock enterprises, shady and fraudulent companies continued throughout the nineteenth century to attract the rogue as principal, and the gullible as investor. Insurance provided a particularly fruitful field of operations. No working capital was required, and there was no expense incurred on plant or machinery. Furthermore, furious competition between the many offices that had come into existence since the company mania of the twenties unleashed on the public a campaign of none-too-ethical advertising, in which unprincipled adventurers eagerly participated. J. R. McCulloch, in *A Dictionary of Commerce* (1845), alerted his readers to the prevailing tone of life insurance transactions:

> A great relaxation has taken place, even in the most respectable offices, as to the selection of lives. And the advertisements daily appearing in the newspapers, and the practices known to be resorted to in different quarters to procure business, ought to make every prudent individual consider well what he is about before he decides upon the office with which he is to insure. Attractive statements . . . ought not to go for much. Life insurance is the most deceptive of businesses; and offices may for a long time have all the appearance

of prosperity, which are, notwithstanding, established on a very insecure foundation. (p. 725.)[1]

As always, the burden of choosing an honest company rested upon the rather bemused and inadequately protected public. A man looking for such an office in 1836 would have been attracted by the prospectuses of five brand-new companies established in that year, the Liverpool and London and Globe, the Northern Assurance, the Hand in Hand, the Legal and General, and the Independent and West Middlesex Fire and Life Assurance Company. The first four were houses of the highest repute, still surviving and flourishing, either independently or in amalgamation; the fifth proved to be the most gigantic and impudent insurance fraud ever perpetrated.

The company was launched by a certain Thomas Knowles, an ill-educated former smuggler and shoemaker, assisted by another ex-smuggler and footman, William Hole. Other active directors were Hole's brother-in-law, William Edward Taylor, a bell-hanger by trade, and William Wilson, a domestic servant. This precious gang was able to open an impressive office in Baker Street, and, within a very short time, to extend its business to Edinburgh, Glasgow, and Dublin. For three years or more the national and local newspapers carried the company's impressive advertisements, of which the following is a typical example:

Immediate Benefits Offered to the Public—Life Annuity Rates, calculated on equitable principles . . . INDEPENDENT AND WEST MIDDLESEX ASSURANCE COMPANY, opposite the Bazaar, Baker Street, Portman Square, London; St. David's Street, Edinburgh; Ingram Street, Glasgow; and Sackville Street, Dublin; established and empowered under the several Acts of Parliament.

. . . Capital, One Million.

By Order of the Board.

Resident Secretary—Mr. William Hole.

Bankers— { The Bank of England.
Bank of Ireland.
Western Bank of Scotland. }

[1] Many insurance companies formed at this period were honest, even altruistic, undertakings, and a considerable number have survived. Such companies as the Guardian (1821), North British and Mercantile (1823), and the Scottish Equitable (1840), were always untainted by 'shadiness' or temptations to speculate, and there were many more. A complete list of the companies appears as the Appendix to H. Withers's *Pioneers of British Life Assurance* (1951).

This notice appeared on page 2 of *The Times* on Thursday, 3 January 1839, a few months before the gigantic swindle was fully revealed. One should notice the meaningless phrase 'established and empowered under the several Acts of Parliament'. In 1836, when the company was established, it was, in terms of the Companies Act of 1825, merely a notional partnership amenable to common law. A further Act of 1837 gave such companies the opportunity to be registered if they so desired. By leaving the choice of registration to the promoter, the Government allowed him the choice of working either openly under rudimentary control, or as secretly and as irresponsibly as he wished. These Acts conferred no powers of establishment, and certainly did not 'empower' the West Middlesex board to do anything. In 1836 it was simply a common-law joint-stock partnership, and it assuredly did not choose to register itself under the Act of 1837. Thus it was possible for the auditor to double as porter, and for a sixteen-year-old floor-sweeper and errand boy to sign annuity deeds for thousands of pounds.

The company furnished its clients with a prospectus that offered outrageously liberal terms,[2] its premiums being 30 per cent lower than those of any other office; and for three years money came rolling in to the coffers of the firm. Some policies falling due were honoured, others were convincingly disputed, but nothing, it seemed, could stem the tide of eager would-be annuitants.[3] The principals spent lavishly, keeping carriage- and saddle-horses, and staffs of liveried servants, and in the manner of the day entertained 'Society' with musical soirées and sumptuous dinners.

Informed opinion was well aware of the unsatisfactory nature of the insurance business. A leading article in *The Times*, in September 1838, drew attention to some of the prevailing abuses. It described 'the very

[2] There is an amusing parody of a company prospectus in *Punch* vol. i, p. 81 (1840). The 'prospectus', for the 'New Grand National and Universal Steam Insurance, Railroad Accident, and Partial Mutilation Provident Society, Capital, Five Hundred Millions', combines the prevailing fears of railway travel with the florid style of current commercial prospectuses. Prospectuses for joint-stock enterprises at this period were often merely fantasies, and the Act of 1844 stated that the promoters of a company were thenceforward to be responsible for any statements made in their prospectuses.

[3] 'The terms were very attractive; there is always a large ignorant class ready and willing to be duped; and the business went on swimmingly . . . If a man wanted to insure his life, there was no great difficulty about his health . . . The poor and less intelligent . . . took their spare cash and invested it in the West Middlesex. Rich men were not less dazzled by the golden promises . . .'. J. F. Francis, *Annals, Anecdotes and Legends of Life Assurance* (1853), p. 229. My account of the West Middlesex fraud is derived largely from this work.

considerable variations existing between the rates of different assurance companies', which had 'latterly become the subject of general remark'. So keen was the competition that well-established companies would make offers to job contracts for clients on much lower terms than those offered in their own tables of standard rates. Such practices made it possible to realize why the public were willing to accept policies from the West Middlesex without too much suspicion of the low premiums offered. *The Times* leader condemned companies like the West Middlesex, and with them the abortive company legislation of 1837 which helped them to flourish.[4]

By 1839, the responsible insurance houses were fully persuaded that the West Middlesex operation was a monstrous fraud. No move was, however, made by these companies to institute an investigation. Had they moved, they could have been accused of attacking an enterprising competitor. More important, they would have been liable to actions for libel, as, in law, the West Middlesex had committed no indictable offence. In London, however, a campaign against the company and its directors was opened by a popular and well-known City figure, J. T. Barber Beaumont.[5]

Undeterred by the laws of libel and slander, Beaumont stated in public that the West Middlesex was fraudulent, and bore the cost of an action brought against him by the company. This action, in the event, was abandoned, as the train of events started by Beaumont gathered momentum. However, Knowles, sensing that public confidence would soon collapse, wrote to Sir John Rae Reid, Governor of the Bank of England, offering him a directorship. Reid

[4] '. . . the absence of any legislative interference . . . against the possible projects of unprincipled adventurers is greatly to be lamented. As the law stands now, no more is wanted to form an assurance company than a batch of directors, whose names may be better known than their property, with a subscribed or nominal capital, the larger in amount the more showy; these, with advertisements and placards have heretofore, and might again, suffice to constitute the whole capital stock of a company . . .' — *The Times*, 7 Sept. 1838, p. 5. (City Intelligence). The description of 'unprincipled adventurers' seems to refer to the by then known methods of the West Middlesex. The situation had not changed a year later. (See the letter, signed 'X.Y.', *The Times*, 12 Nov. 1839, p. 3.)

[5] For details of the colourful life of this insurance broker see *DNB*, and Walford, *Insurance Cyclopaedia*, i, pp. 261–2. He died in 1841. Both Francis (op. cit. p. 231), and Withers (op. cit., p. 61), state that Beaumont wrote a letter to *The Times*, exposing the West Middlesex, in 1839. I cannot trace this letter, which seems not to have been printed by the editor. For Sir John Rae Reid's letter to Mackenzie, see *The Scotch Reformers' Gazette* for Saturday, 2 March 1839. Part of the letter is indirectly quoted in Francis, *Annals, Anecdotes and Legends*, p. 234.

contemptuously refused, and ordered Knowles to remove the West Middlesex accounts from the Bank.

It was the Scottish journalist Peter Mackenzie who continued Beaumont's work of exposure in the pages of his *Scotch Reformers' Gazette*. Having ascertained from Sir John Rae Reid that the Bank had severed all connection with the West Middlesex, he wrote to the Duke of Wellington, whom the trickers had claimed as a patron of the company. The Duke replied that there was no truth in this statement, and that his name had been used by 'a gang of swindlers'.[6]

Convinced now that he must inform the public of the true state of affairs, Mackenzie launched his remarkable series of attacks in the *Scotch Reformers' Gazette*. A leading article headed 'Exposure' appeared on Saturday, 2 March 1839, in which he damned the West Middlesex as 'a spurious insurance company hatched in London two years ago' by 'a parcel of tricksters in London'. Despite the immediate commencement of an action for libel by Knowles and his associates, Mackenzie returned to the attack on Saturday, 9 March, revealing how 'the spacious office of the *tricksters'* in Ingram Street, Glasgow, had been besieged by a frantic crowd, and how the principals had attempted to refute Mackenzie's accusations by printing a statement in the *Glasgow Herald* on Monday, 4 March, and a further frantic refutation in the same paper on Friday, 8 March.

In his devastating attack of 9 March, Mackenzie tore the refutation to shreds, as 'the lame and impudent production' of 'a company of London *swindlers'*. Referring to their claim in their advertisements to be 'empowered under the several Acts of Parliament', he wrote:

[6] I am very grateful to Mr W. A. G. Alison, FLA, City Librarian, and to the staff of the Mitchell Library, Glasgow, for locating the various articles from the *Scotch Reformers' Gazette*, and for sending me photocopies of them. Many joint-stock companies advertised prominent aristocrats as either patrons or members of the Board, and the practice is by no means extinct. The West Middlesex's company director listed the names of Drummond, Perkins, Smith, Price, and Lloyd as members of the Board, but the initials accompanying these surnames were not those of these celebrated bankers: the whole list was an impudent fiction. Francis (*Annals, Anecdotes and Legends*, p. 235) reproduces part of the Duke of Wellington's reply to Mackenzie, in which the Duke refers to 'a gang of swindlers'. There is no trace of this correspondence at Stratfield Saye. I am very grateful to His Grace the Duke of Wellington, and to his archivist, Miss Joan Wilson, who undertook a search of the Wellington Archive on my behalf. Miss Wilson informs me that the Duke was the subject of a number of fraudulent attempts to connect his name with various businesses, and that the letter quoted by Francis had probably been written by a secretary.

do they presume to vindicate that character which . . . they made the public to believe they possessed by virtue of some Charter of Incorporation, or by the authority of some special Act of Parliament in their favour?

Would-be clients could well have read these interpretations into the phrase, the impression of wealth and status being enhanced by the cachet of a Royal Charter, and the vast expense of a private Bill.

Continuing his attack, Mackenzie poured scorn on their much-vaunted capital: 'But their capital of *one million of money*—where is it? How—when—where was it subscribed—or where is the whole or any part of it deposited?' There could be no answer to these questions, as the capital of 'one million' did not exist. Issue after issue of Mackenzie's paper continued this ruthless exposure, and Sir John Rae Reid in London felt emboldened to inform the Lord Mayor, Sir Peter Laurie, that the West Middlesex directors were the greatest swindlers ever known in London. Laurie now prepared to move against the company, but before he could act the Baker Street premises were hurriedly vacated, and the perpetrators of this gigantic fraud absconded. The loss to the public in stolen premium-money was in the region of £250,000.

In the early forties, when the lessons, financial and moral, to be learnt from this famous swindle still retained their force, Dickens and Thackeray created fictional insurance companies which have interesting affinities with the West Middlesex, though each author makes rather different use of this particular commercial stimulus to literary composition.

From September to December 1841, *Fraser's Magazine* serialized Thackeray's *The History of Samuel Titmarsh and the Great Hoggarty Diamond*.[7] This work chronicles the doings of Mr. Brough, ostensibly a Turkey merchant, but primarily a company promoter with interests in five hundred firms. He it is who brings into being the 'Independent West Diddlesex Fire and Life Assurance Company'. The name of this company quite obviously echoes that of the real-life West Middlesex, which had crashed some two years before the serial appeared, but, like some other novelists of the era, Thackeray chooses to place his tale of chicanery in the early twenties, 'when, as readers may remember, there was a great mania in the City of London for establishing companies of all sorts, by which many people made pretty fortunes.' (p. 13.)

[7] It appeared in volume form in 1849. Page references in the text are to the edition of this work in The Oxford Thackeray, ed. George Saintsbury, 1909.

The bulk of Brough's operations occur during 1823, and this dating means that the West Diddlesex functions before the Companies Act of 1825, which had rendered joint-stock enterprises amenable to common law. As a notional partnership only, Brough's Diddlesex could neither sue nor be sued in the name of its officers. Thus Thackeray chooses to place his insurance company in a commercial world that was rather different from that of the late thirties.

Thackeray tells us nothing of Brough's origins: he is from the start the complete director, Member of Parliament for Rottenborough, and owner of a splendid villa at Fulham. There is no suggestion of an obscure origin, and there is very little in the story to suggest that Thackeray has in any way consciously modelled his West Diddlesex Company and its principal on the West Middlesex and its directors. The office is staffed by an encouragingly realistic set of managers and clerks, and the company secretary, Roundhand, is a genuine actuary.

For most of the story Roundhand is a circumspect supporter of the company, but towards the end he quarrels with Brough. (A quarrel between the principal and the secretary was one important cause of the fall of the West Middlesex. Each accused the other of misappropriation of funds, and William Hole, the Resident Secretary, quitted Baker Street, to add his voice to Mackenzie's in damning the company he had helped to create.[8]) Roundhand's quarrel with Brough—again over money—leads to his leaving the West Diddlesex. Having withdrawn his own money from the concern, he disposes of his shares 'to a very good profit', and thereupon spreads adverse rumours about his erstwhile employer. Roundhand, like Hole, is fully aware of his principal's malpractices, and Thackeray gives us a glimpse of this awareness in the following words of Brough:

'And oh, Roundhand!' continued our Governor, 'draw a cheque for 700, will you. Come, don't stare, man, I'm not going to run away! That's right—seven hundred and ninety, say, while you're about it! Our board meets on Saturday, and never fear, I'll account for it to them!' (Ch. 17.)

Brough's activities as a company promoter of the 1820s are given equal prominence with his insurance business: his West Diddlesex is just one of some five hundred typical 'bubble' companies with which he is

[8] Francis reproduces a letter sent by Hole to Knowles in which he writes to his erstwhile friend: 'Thou art a scoundrel . . . you damned perjured scoundrel!—you base wretch!', and threatens to 'expose your villainy to Mackenzie and others' (Francis, *Annals, Anecdotes and Legends*, p. 240).

linked. Thackeray carefully details his business methods, showing for instance how in January 1823 he declared a half-yearly dividend of 41 per cent, a return that could only have been paid out of capital.[9] Operating before 1825, there is no significant company law for Brough to contravene, and it is as well to remember that neither aggrieved shareholders nor duped investors could seek criminal restitution, as technically both parties were *in pari delicto*.[10]

Brough himself is to some extent a victim of the imperfect state of company law. Needing a fresh issue for his 'Patent Pump Company' when its shares stood at 65, Brough and his directors called upon the shareholders to come up with more money. They simply refused to pay, and there was nothing in law that Brough could do about it. The bubble burst when Brough's associate companies collapsed, one after the other. The process occupies three pages of rapid narrative in which facts familiar to contemporary readers, who would have remembered the traumatic events of the 1825 crisis, are revealed.

> About this time, in the beginning of 1824, the Jamaica Ginger Beer Company shut up shop . . . The Patent Pump shares were down to £15 upon a paid-up capital of £65 . . . the Muff and Tippet Company shut up, after swallowing a capital of £300,000 . . . (p. 95.)

At this stage malversation, rather than mismanagement, is suggested, and when it is rumoured that Roundhand is about to risk proceedings against Brough, the latter prudently abandons his fallen empire to seek refuge in France.[11]

It is clear that Thackeray, like Miss Mulock, Reade, and others, looked to the 1825 era to furnish him with material for a shady company promoter, whom he needed to manipulate the plot of his novella, and to provide him with the opportunity to castigate the

[9] In Bulwer's *The Caxtons* (1849), the irrepressible Uncle Jack pays the shareholders of his 'Coal Company' 20 per cent dividend a year out of their own capital; the company never earns a penny.

[10] That is, lying beyond the law to which they appeal. Victims frequently cut their losses, or even compounded for part restitution by hiring a member of the detective police to trace the defaulter, and arrange a compromise.

[11] The loss of the enormous capital of £300,000 in the 'Muff and Tippet Company' shows how Brough must have regarded the company's capital as his own private property, and at his examination in bankruptcy it is proved that he had been criminally irresponsible in his use of invested money. Part of this money would have gone in paying 'dividends', and the rest into other private speculations.

foolishness and greed of some of the victims.[12] The name of his insurance company echoes that of the West Middlesex, but the story is firmly centred in the familiar ambience of the great Company Mania.

Of much greater interest is Dickens's depiction of a fraudulent insurance promoter in *Martin Chuzzlewit*. In this novel, which was issued in monthly parts from January 1843 to July 1844, we meet 'Tigg Montague, Esquire (of Pall Mall and Bengal)', proprietor of 'The Anglo-Bengalee Disinterested Loan and Life Assurance Company'. Dickens was fully aware of the events of 1825, and I have shown elsewhere how he used them to advantage in *Nicholas Nickleby*.[13] The action of *Martin Chuzzlewit*, however, is contemporary, if we are to judge from the reference to Queen Victoria in Chapter 21, and although the name of Dickens's company does not imitate that of the West Middlesex, it has many interesting affinities with it, suggesting that the details of the notorious swindle were in the author's mind when he created the Anglo-Bengalee.

Tigg's Anglo-Bengalee, like the West Middlesex, carried on its nefarious business during the time that the largely ineffectual Company Act of 1837 was in force. In Chapter 27 we are informed by David Crimple of Tigg's 'property in Bengal being amenable to all claims upon the company', which suggests a voluntary registration under the Act, the (non-existent) estate thus satisfying the stipulations of Section 3 of the Act, which would allow the firm to sue or be sued in the name of one or two of its officers. One cannot, of course, prove the point either way. The avowed intentions and progress of Tigg's company suggest complete secrecy, and a reliance upon the stipulations of the Act of 1825.

Thackeray's Brough was provided with no antecedents, but Tigg's origins are clearly parallel to those of the real-life Knowles and Hole. Tigg is a product of 'low life' and obscurity, who uses an amoral cunning to transform himself into Tigg Montague, a particularly shining example of the 'swell mob'.[14] In the early chapters of

[12] The Commissioner in Bankruptcy tells young Titmarsh: '. . . if you had not been so eager after gain, I think you would not have allowed yourself to be deceived' (p. 117.)

[13] N. Russell, '*Nicholas Nickleby* and the Commercial Crisis of 1825', *The Dickensian*, Autumn 1981, pp. 144–50.

[14] 'There remained in the shadow that George IV continued to cast long after he had disappeared from mortal view . . . a body known as the "Swell Mob", of whom Sir Mulberry Hawk, Montague Tigg, Captain Costigan, and others are the best known representatives in fiction.' R. H. Mottram, 'Town Life', in *Early Victorian England* (1934) i, p. 184.

Martin Chuzzlewit he is depicted as a frequenter of taverns and pawnshops, dressed in dirty rags; in effect, a tramp. He tells Pecksniff and Tom Pinch outrageous lies about his own and his father's prowess in war, in which one detects a cunning and mendacity born of the poverty of the streets. Dickens here originates his speculator in the same social stratum as that of Knowles and Hole, men who had found in honesty only failure and deprivation. Both men had held menial posts, and both had engaged in smuggling before turning their native talents to fraud.

Although Thackeray tells us little of Brough's personal appearance and dress, Dickens describes his transmuted Tigg in detail, with an underlying purpose that is largely lost on the modern reader.

> He had a world of jet-black shining hair upon his head, upon his cheeks, upon his chin, upon his upper lip. His clothes, symetrically made, were of the newest fashion and the costliest kind. Flowers of gold and blue, and green and blushing red, were on his waistcoat; precious chains and jewels sparkled on his breast; his fingers, clogged with brilliant rings, were as unwieldy as summer flies but newly rescued from a honeypot. The daylight mantled in his gleaming hat and boots as in a polished glass. (Ch. 27.)

This closely-observed description conforms to a then-prevailing image of an untrustworthy upstart, and this is how Dickens's readers would have taken it. G. H. Lewes, in his review of the fifth edition of Disraeli's *Coningsby* in 1849, produces a penetrating contrast between the dress of a 'gentleman' and that of an upstart social parasite. The gentleman, by some subtle chemistry, reveals that he is such, even if 'dressed to drive tailors to despair'. Lewes then points the contrast in these words:

> Compare such a man with one of those 'striking' specimens of modern society, who, with radiant waistcoat, resplendent jewellery, and well-oiled whiskers, lounges through the public promenades 'the observed of all observers'; *him* you do not mistake for a gentleman.[15]

It may be assumed, from the close similarity of the expressions used by Dickens and Lewes, that both are describing a familiar contemporary type, recognized instantly by the judicious as undeserving of respect. Lewes, who disliked Disraeli's work, declared that *Coningsby* was like a jewelled ring sparkling on a dirty finger,

[15] [G. H. Lewes.] 'Coningsby', *British Quarterly Review* No. 10 (1849), p. 120.

and Francis describes how Knowles and Hole ordered their fellow-directors to dress on board days in their 'Sunday best', 'to place brooches in their dirty shirts, and rings on their clumsy fingers.'[16] This functional description shows Dickens's concern to indicate a social basis for his fraudulent company which is at least similar to that of the West Middlesex.

Tigg's business associates belong to the same dubious world as those of Thomas Knowles. The ex-smuggler Knowles appointed his former confederate 'resident secretary' at the main office of the West Middlesex in Baker Street; Tigg goes into partnership with David Crimple the pawnbroker, who, in the same way, becomes 'Secretary and resident Director' of the Anglo-Bengalee at offices in 'a new street in the City'. Residence of the secretary at the principal office created confidence in the public mind. William Hole's successor as secretary of the West Middlesex claimed to be resident at Baker Street in March 1839, although he most certainly was not. No doubt Mr Crimple's continued residence at the office was a similar 'polite fiction', as he very easily absconded with all the loot.

Francis tells us that Knowles concocted his scheme as 'the best mode of supplying his necessities';[17] in similar fashion Dickens informs us that both Tigg and Crimple have 'come into a few pounds', which they can put together 'to furnish an office and make a show'. In a business that required no plant or fixed assets, the external show, seeming to guarantee the 'golden promises', was essential for the imposture to be successful. Referring to the West Middlesex's much-vaunted 'One Million', Francis drily remarked: 'The chief capital demanded by such an undertaking on the part of the proprietor, was unbounded impudence; and on that of the public, unbounded credulity.'[18] Tigg's company can outdo even Knowles's in 'unbounded impudence' when it comes to stating its capital: 'A figure of two, and as many oughts after it as the printer can get into the same line.'[19]

16 Francis, *Annals, Anecdotes and Legends*, p. 227. Looking beyond Francis's contempt for the West Middlesex swindlers, one can see how they must have conformed in dress with Tigg and with G. H. Lewes's character study. There seems to have been a prevailing image for the parvenu, that of dirty or clumsy hands covered with sparkling rings.

17 Francis, *Annals, Anecdotes and Legends*, p. 225.

18 Ibid., p. 226.

19 *Martin Chuzzlewit*, Ch. 27.

Tigg and Crimple make their 'show' by furnishing impressive offices in the City. Beyond an outer office, 'where some humble clients were transacting business', there lies

> an awful chamber, labelled Board room—the door of which sanctuary immediately closed, and screened the great capitalist from vulgar eyes . . . The chairman having taken his seat with great solemnity, the secretary supported him on his right hand, and the porter stood bolt upright behind them . . . This was the board—everything else was a light-hearted little fiction. (Ch. 27.)

These words show clearly that the Anglo-Bengalee has been established in the same manner as had the West Middlesex. At Baker Street, the boardroom was usually empty, as the only real directors were Knowles and Hole. It was shown earlier how many of their 'directors' were mere fictions, and how a passable list had been compiled from the addresses of errand-boys and servants. Tigg is in the same position: he has an imposing boardroom, but nobody to fill it. We notice the presence of the porter in the boardroom. The West Middlesex porter had doubled as auditor, and it was perhaps with him in mind that Dickens created the splendid porter of the Anglo-Bengalee, Bullamy, leaving us to decide 'whether he was a deep rogue or stately simpleton'. Certainly Gates, Brough's porter at the West Diddlesex, was a stately simpleton, as he became his master's victim, losing his modest life-savings when the company crashed.

Another similarity between the Anglo-Bengalee and the West Middlesex was that both companies made much use of advertising, a feature not mentioned by Thackeray. The West Middlesex was renowned for the success of its unremitting advertising campaign. The company circulated its prospectuses to remote country districts. It advertised not only in the daily and weekly newspapers but in the monthly and quarterly journals, and, as Francis tells us, 'the walls of provincial towns absolutely blazed with their attractive terms' (p. 231). Mackenzie gives us a brief but vivid description of their Glasgow premises, 'with placards out upon his windows, "Immediate Benefits", and so forth . . .'.[20]

Dickens compresses this aspect of the insurance business into a long and breathless paragraph, in which he shows the company's name emblazoned on every article, movable and immovable, in the office:

[20] The *Scotch Reformers' Gazette*, 6 Apr. 1839.

Publicity! Why, Anglo-Bengalee Disinterested Loan and Life Assurance Company is painted on the very coal-scuttles. It is repeated at every turn until the eyes are dazzled with it, and the head is giddy . . . it is repeated twenty times in every circular and public notice . . .[21]

Dickens creates one interesting functionary not mentioned by Thackeray, but found attached to all Life Offices of the day. This was the medical officer, in *Martin Chuzzlewit* the celebrated Dr Jobling. These medical officers seem to have been in the main a dubious class of men, to judge from remarks made by Francis and McCulloch. Francis remarked that 'If a man wanted to insure his life, there was no great difficulty about his health', and McCulloch declared: 'Since the competition among the different offices became so very keen as it has been of late years, there are but few lives so bad that they will not be taken by one office or another.'[22]

Regular medical officers, paid by the board, were an established feature of the companies by the thirties, and their examination of prospective clients was based on the standard questionnaires that had to be answered.[23] Jobling, described as 'a paid (and well-paid) functionary', clearly saw few difficulties about passing clients as fit. He is, in fact, a genteel tout for the company, and on the occasion when he visits Tigg at the office, he brings him four new policies, for which he receives commission. Doubtless there were many such 'retained physicians', not over-nice in their diagnoses.[24]

The Anglo-Bengalee is one of the few business enterprises in Dickens's novels to be delineated in some detail, and its methods of business are clearly explained in Chapter 27 of *Martin Chuzzlewit*. New policies are granted on a vast scale, liberal allowances are afforded to solicitors, and loans are made available, coupled to life-assurance, bonded sureties, and the introduction of new business. One may see in these methods of business an echo of the West Middlesex swindle. The 'immediate benefits' offered to the public in their advertisements

[21] There is a little touch of irony in the posting of placards for the apprehension of Crimple in Ch. 49. He is 'advertised for, with a reward, upon the walls'.

[22] Francis, op. cit., p. 229; McCulloch, op. cit. p. 725.

[23] A medical or non-medical referee was to be provided, who would state whether the annuitant had ever suffered from consumption, gout, asthma, smallpox, whooping-cough, etc. (McCulloch, p. 728b).

[24] More expert opinion was forthcoming from the Clerical, Medical and General (1824), the Board of which was composed mainly of distinguished medical men. (Withers, *Pioneers of Life Assurance*, pp. 55–7). Appearance before the Board itself, not just before the medical officer, lingered on with some companies into the 1890s.

were often loans, granted to the applicant in return for his insuring his life with the company, and persuading two friends to guarantee his bona fides by themselves effecting an assurance bond. The applicant's first premium would be deducted from the loan or, if agreeable, from the friends' bond-money. This was a very effective—and quite legal—method of gaining a number of clients at one stroke.[25] Tigg offers his hapless clients the same lure of 'immediate benefits', which leads to the taking of further premiums, and such is Tigg's genius for fraud that even the cunning Jonas Chuzzlewit falls into the trap.

Both Tigg and Brough are aware of the chief danger of the early stages of a Life Office's growth, and it is interesting that Dickens and Thackeray express the danger in very similar ways:

> life insurance companies go on excellently for a year or two after their establishment, but it is much more difficult to make them profitable when the assured parties begin to die. (*Hoggarty Diamond*, p. 100.)

> 'But when they begin to fall in', observed Jonas. 'It's all very well, while the office is young, but when the policies begin to die; that's what I am thinking of.' (*Martin Chuzzlewit*, Ch. 27.)

Brough had no desire to experience a crash, and it was just unfortunate for him that the collapse of the bubble companies demolished that public confidence necessary for survival, and so secured his ruin. Tigg has no such cause for worry. When serious calls are made upon the company, he will not be there to answer them. '"Whenever they should chance to fall in heavily . . . then—" he finished the sentence in so low a whisper that only one disconnected word was audible, and that imperfectly. But it sounded like "Bolt".' (Ch. 27.)

This examination of the Anglo-Bengalee has shown how closely Dickens's fictional company parallels the West Middlesex. If Dickens had not that particular swindle in mind at the time of composition, then at least he brought to his writing a very well-informed knowledge of then-current commercial insurance practice. This particular branch of commerce brought much drama in real life and, as we have seen, provided admirable material for the stuff of Victorian fiction.

[25] Various means were employed by these fraudulent insurance companies to attract custom. Agents, whose usual commission was 5 per cent on policies effected, were offered 25 per cent by the West Middlesex. Such was the impressive scale of the fraud that many agents of genuine repute never questioned the bona fides of a company that could offer such a high figure (Francis, *Annals, Anecdotes and Legends*, p. 229).

However, in neither of the works examined is the author content to stand outside his creation: Dickens and Thackeray part company with Francis in staying with their promoters after the demise of their companies has been clearly signalled. Thackeray positively warms to Brough, showing how, like his wretched dupes, he loses everything, and exists in a garret in Boulogne, sustained by his brave and faithful wife. He describes him as a man of 'undaunted courage', in whom there must be some good, as his family refuse to forsake him. These final images of Brough form a rather startling contrast to the bland religious hypocrite with the biblical manner and a talent for coarse bullying depicted throughout the tale before the crash, and one suspects that this is a typically generous volte-face from Thackeray, unwilling, towards the close of his work, to leave a lively character entirely blackened.

Whereas Thackeray provides sufficient material for a credible explanation of the West Diddlesex's downfall to be adduced, Dickens, towards the end of his novel, becomes impatient with his commercial creation, as he begins to remould Tigg into a new, more spectacular image. Excited by a new dimension of his creative instinct, he becomes careless of the Anglo-Bengalee, just as, in the later novel *Dombey and Son*, he quickly loses interest in Dombey as a merchant.[26] Thus the crash of the Anglo-Bengalee is revealed at third hand, in an abrupt, laconic sentence, for which there has been no detailed preparation: 'Their office is a smash, a swindle altogether' (Ch. 49).

Tigg is not destined by his creator to be merely a swindling insurance promoter. He is also to be a blackmailer, who is murdered by his victim, Jonas Chuzzlewit. Dickens involves these two men in a deadly, melodramatic relationship from Tigg's first appearance as a promoter in Chapter 27, from which point the bursting of the bubble becomes of decidedly secondary importance.

Tigg is an unevenly developed character, fulfilling several different roles throughout the novel, and at times one wonders what Dickens meant to do with him. In the opening chapters he is described as being dirty, mean, and slinking, yet despite these qualities he is developed to some extent as a fruitful source of humour, having affinities with the earlier picaresque creation, Mr Jingle. His various encounters with

[26] For the almost total lack of 'dealings' with the firm of Dombey & Son, suggesting that Dickens is preoccupied with other aspects of the novel, see in particular Butt and Tillotson, *Dickens at Work* (1957), p. 109, and R. H. Dabney, *Love and Property in the Novels of Dickens* (1967), p. 55.

young Martin, Mark Tapley, and Tom Pinch are designed to emphasize the sterling qualities of those characters, but one still remembers Tigg's outrageously amusing posturings.

Then, in Chapter 27, this sponging mendicant reappears, after a long absence from the novel, as a fully-fledged man of business, 'the great speculator and capitalist' as Dickens calls him. Most of his earlier bombast and skittishness has gone, and he is shown as a shrewd entrepreneur. His language is more controlled and purposeful, and, as we have seen, he has become a master of insurance practice, a realistic echo of Knowles and Hole.

In this new guise he becomes involved with the hypocrite Pecksniff, and it becomes clear that Tigg's rise will ensure Pecksniff's fall. This is a meeting of two consummate rogues, both pursuing their own selfish and mercenary ends by different means, although the arch-hypocrite will fall an easy victim to the amoral parasite, who now feeds off society at large, not merely off a single family.

In Chapter 38 Dickens involves Tigg and Jonas Chuzzlewit in the preliminaries of a Dickensian murder, and once again Tigg as a character changes direction. He becomes a potential victim, as his attempts at blackmail drive Jonas to plan his murder. Dickens's well-known preoccupation with murderers and their victims begins to take control not only of the novel's structure but of Dickens's consistency in character-depiction.[27] Both Jonas, the Manchester-goods merchant, and Tigg, the insurance principal, are now lifted out and away from the main theme and assigned a drama of their own, which is acted out against an atmospheric backdrop of wild countryside and cosmic storm. The fated and fatal bond between the two men is cemented further by the skilful interweaving of their dreams.[28]

By Chapter 47 Tigg is fully metamorphosed into the marked victim of 'bastard murder'. Any didactic or moral interest in the fate of Tigg the insurance promoter has been utterly lost as this new Abel goes forth to meet his Cain. For Tigg there will be no flight, gaol, or suicide. Dickens just hints at the possibility of repentance—'it may be that the evening whispered to his conscience'—before he sends him to

[27] R. H. Dabney has called this controlling fascination 'the Bill Sikes syndrome', and shows how it causes Dickens to alter Jonas's character quite radically. See also E. B. Benjamin's attempt to define a unifying structure for the novel in 'The Structure of "Martin Chuzzlewit"' in *Philological Quarterly*, No. 34 (1955), pp. 39–47.

[28] See in particular the brilliant essay by J. Brogunier, 'The Dreams of Montague Tigg and Jonas Chuzzlewit', *The Dickensian* No. 58 (1962), pp. 165–70.

his doom. Tigg's role in the novel has reached its final change, this time a change almost beyond recognition.[29]

By the time, therefore, that Dickens reaches his concluding chapter, he has behind him the long movement of the story in which Tigg reaches his final transmutation, and winding up the affairs of Tigg and Jonas as insurance swindlers is of decidedly secondary importance or concern to him. Nevertheless he must, in this final chapter, tie up the loose ends of his insurance theme, and he leaves it to his *deus ex machina*, old Martin, to reveal all to the widowed Mercy Chuzzlewit.

> Your late husband's estate, if not wasted by the confession of a large debt to the broken office (which document, being useless to the runaways, has been sent over to England by them—not so much for the sake of the creditors as for the gratification of their dislike to him, whom they suppose to be still living), will be seized upon by law; for it is not exempt, as I learn, from the claims of those who have suffered by the fraud in which he was engaged. (Ch. 54.)

It would appear that 'those who have suffered from the fraud' have sufficient funds to institute proceedings for restitution against Jonas's estate, using for this purpose a 'confession of a large debt' owed by Jonas to the 'broken office'. There is some confusion in all this. Who, for instance, are 'the runaways'? We know that Crimple has 'bolted', presumably to the continent; we also know that he and Tigg were the only directors, the rest of the board being 'a light-hearted little fiction'. Yet Crimple has multiplied into 'the runaways', and there is the suggestion of a feud between 'them' and Jonas which is nowhere else even hinted at in the novel. If Jonas had incurred a debt with the office, it must have been in the form of a loan, so that he, in effect, was a victim. By what means, or for what reason, Jonas was induced to 'confess' the debt, we are not told.

Further, Dickens seems to have forgotten that Jonas was a Manchester warehouseman, a distributive merchant for cotton goods. As such, he would normally have raised loans on the security of his business by means of the usual inland bills of exchange. The 'runaways' could have benefited themselves by securing acceptance of such normal bills, and then having them discounted for gold at a

[29] So completely has Dickens lost sight both of the comic rogue and the swindling financier, that he enshrines him in this romantic and heavily symbolic vignette: 'The glory of the departing sun was on his face. The music of the birds was in his ears. Sweet wild flowers bloomed about him. Thatched roofs of poor men's homes were in the distance: and an old gray spire, surmounted by a cross, rose up between him and the coming night.' (Ch. 47.)

bank or discount-house. The purpose and nature of this loan, and why Jonas should have 'confessed' it, must remain a perplexing mystery. We see here some hasty plot-construction, and the results of having to tie up a tiresome loose end. Crimple, incidentally, is left to enjoy his ill-gotten gains in peace.

Dickens himself was prudent enough to insure with companies of impeccable reputation. On Tuesday, 9 January 1838, when the West Middlesex was in mid-career, he presented himself before the Board of the Sun Life Assurance Society, established in 1810, for the customary medical examination. Dr. Jobling was not in evidence, and Dickens was examined by Henry H. Southey, MD, FRS, brother of the poet, and at one time physician to George IV. The Board were concerned that Dickens worked too hard, but seem to have granted him a Life policy.[30]

In November 1841 Dickens effected a policy with the Eagle Life Assurance Company (1807). The secretary and actuary of this company, H. P. Smith, was largely instrumental in the apprehension of Thomas Griffiths Wainewright, the poisoner and insurance defrauder. Wainewright, an associate of Hood, De Quincey, and Charles Lamb, was an art critic of no mean talent, but from early days had shown a propensity for crime. Left an annuity of £200, he forged an order upon the Bank of England to obtain the capital sum, an act for which he was to pay dearly in later years. His forgery was committed in 1826, and two years later he went to live with his bachelor uncle, George Griffiths, who died suddenly and left Wainewright all his property. In 1830 his mother-in-law, who had objected to his insuring the life of her unmarried daughter, died as suddenly, and immediately afterwards the daughter, Helen, now insured for several thousand pounds, also died. It was Dickens's friend, H. P. Smith, who disputed Wainewright's claims upon the Eagle Life Assurance Company. Wainewright prudently fled to Boulogne in 1831 but, deeming it safe to return six years later, was arrested for his forgery of 1826 against the Bank of England. There is no doubt that he poisoned all his relatives, but legal proof was lacking. On the forgery charge he was sentenced to transportation for life, and died in hospital at Hobart in 1852.

Wainewright appears as Varney in Bulwer's *Lucretia* (1846), and Dickens used him as the inspiration for the sinister Julius Slinkton in

[30] See *Pilgrim Edition of the Letters of Charles Dickens*, i (1965), pp. 630–1.

Hunted Down (1859). In this tale Slinkton is made the murderer of one niece and the potential murderer of another, in order to gain their insurance money. He is 'hunted down' by Meltham, 'the young actuary of the "Inestimable"', who is based on H. P. Smith of the Eagle. Slinkton is unmasked by a clever trick and, unable to face retribution, he poisons himself. This melodramatic little tale shows Dickens, as it were, 'on the other side': it stands as a justification of the vast majority of insurance companies, such as the ones with which he himself insured, the honesty of which was axiomatic, and far removed from the perversion of commercial practice seen in the Anglo-Bengalee, and in the Independent and West Middlesex Assurance Company.[31]

Whether honest or fraudulent, the bankers and insurance promoters of the type examined so far were rarely described in superlatives: they were seen as fairly equal rivals in the race to husband the public's money, and none was accorded any special measure of regard as one of the great ones of the earth. There is, however, a stage in the popular imagination where a banker ceases to be such, and becomes a 'financier', a figure of international importance, and a star of great magnitude. The metamorphosis from, say, Drummond, Smith, and Hoare to Baring and Rothschild is to some extent one of degree, but also one of role and function. The financiers and merchant bankers, evoked a wholly different complex of attitudes and responses, both in the public consciousness, and in the individual reactions of the novelists.

[31] See Wainewright's *Essays and Criticisms*, ed. with memoir by W. C. Hazlitt, 1880; T. Seccomb, *Lives of Twelve Bad Men*, 1894; *Pilgrim Letters*, ii (1969), p. 252.

5

Financiers (I): Rothschild, Sidonia, and Mrs Gore's Osalez

Among the 321 private banks existing at the opening of the period, there was a large and important group of London firms that specialized in business connected mainly with foreign trade and investment. These were the merchant bankers, and of their number there were several whose particular business lay in raising loans for the use of governments.[1] Notable among such houses were those of Abraham and Benjamin Goldsmid, Baring Brothers & Co., and N. M. Rothschild & Sons. The Goldsmids, originally merchants, ventured upon stock-broking in 1792, and established the fortunes of their firm by negotiating English loans for Pitt during the long war with France. The great house of Baring was founded by Francis Baring in 1770. Long associated with the East India Company, he too made a fortune in raising loans for the French war: when he died, in 1810, he left £7 million.[2]

Baring's and Goldsmid's continued to flourish throughout the century; but neither firm could seriously rival, either in wealth or in influence, the famous house of N. M. Rothschild & Sons. The Rothschild family, and the banks that its members established and developed, had a wide-ranging influence on the literary depiction of capitalism in the nineteenth century, and for this reason a brief account of the rise and early progress of the house will be given here.[3]

[1] One did not need to be an avowed stockjobber or broker, or, indeed, a member of the Stock Exchange, to finance this type of loan. Private bankers, provided that they commanded sufficient capital, could be what Fox Bourne dryly described as 'stock-jobbers, as it were, on the sly' (H. R. Fox Bourne, *The Romance of Trade* (n.d.), p. 322).

[2] Baring's continued its foreign brokerage throughout the century. In 1890, when Argentina failed to meet its interest payments, Baring's nearly collapsed, with liabilities of £21 million. A consortium, headed by the Bank of England, and including N. M. Rothschild & Sons, averted the disaster, and Baring's was re-established as a limited company.

[3] The standard work on the rise of the Rothschilds is Count Egon Caesar Corti's *Der Aufsteig des Hauses Rothschild* (Leipzig, 1927), cited in this book as *Corti*. I have used the translation, *The Rise of the House of Rothschild* (New York, 1928), by Brian and Beatrix

The Jewish family of Elchanan had dwelt in the Frankfurt ghetto since the sixteenth century. From the small red shield affixed to the frontage of their house, they came in time to be known as Rothschild. Engaged initially in retail trade, they became money-changers in a small way in the early years of the eighteenth century. Meyer Amschel Rothschild, the founder of the family fortunes, was born in 1743. After leaving school, he entered the well-known Hanoverian banking house of Oppenheim, where his innate ability to sense a bargain, and his interest in numismatics, brought him to the attention of Crown Prince William of Hesse, ruler of Hanau, himself an enthusiastic coin collector. Meyer Amschel was wise enough to place many numismatic bargains in the prince's way, and in 1769 William granted him the court title of Crown Agent.[4]

In August 1770 Meyer Amschel married Gutle Schnapper, the daughter of a neighbouring Jewish tradesman. A daughter was born in 1771, followed by three boys: Amschel (1773), Solomon (1774), and Nathan (1775). The family ultimately expanded to ten children, five daughters and five sons. Carl Meyer was born in 1788, and Jacob, known also as James, in 1792. The names of these five sons were to become household words throughout Europe.

In October 1785 Prince William succeeded his father as Landgrave of Hesse-Cassel, William IX. The combination of his own and his father's fortune made him virtually the richest prince in Europe. Rothschild had been employed in a small way by William in discounting bills, and he now petitioned his patron for more regular business. The prince had felt safer with the very old-established firm of Bethmann, but finally in 1794 he agreed that Meyer Amschel should transact regular bill-discounting work for him. This was, in effect, the beginning of the family's rise to greatness, as William was allied by marriage to the royal families of Great Britain and Denmark, in both of which countries he had invested considerable capital. Already, therefore, the Rothschild family were moving away from an insular to an international concept of business.

Lunn; page references are to this edition. Corti is conveniently supplemented by Virginia Cowles, *The Rothschilds, a Family of Fortune* (1973), cited as *Cowles*. Page references are to this edition. See also Lord Rothschild, *The Shadow of a Great Man* (1982).

[4] The title carried the same force as the modern 'By Appointment' granted to traders and manufacturers, with the right to display the Royal Arms. Still effective today, it conferred distinct advantages upon an obscure Jewish trader in eighteenth-century Germany. In 1800 he secured the even more useful title of 'Imperial Crown Agent' from the Roman-German Emperor, Francis II.

The political convulsions following the French Revolution brought further profit to the Rothschilds, especially after the conquest of Holland in 1795, when the Amsterdam exchange collapsed, so that much new business flowed into the Frankfurt money market. Landgrave William, that titled financier, augmented his already great fortune by lending money not only to fellow princes, but to merchants, tradesmen, and artisans. Meyer Amschel had entered into partnership with his two elder sons, Amschel and Solomon, and on several occasions, they had acted as intermediaries in contracting loans made to the Roman-German Emperor by Landgrave William, who was angling for the coveted title of Elector.

The firm's business with England, involving transfer of bills, cash payments, and the conveyance of English merchandise to the whole of Germany via Frankfurt, had increased considerably by the closing years of the century. Accordingly, in 1798 Meyer Amschel's highly gifted third son, Nathan Meyer, then aged twenty-one, suggested that he be allowed to emigrate there to set up as a merchant on his own account, and to act as the Rothschild agent in Britain. The plan was agreed, and Nathan duly arrived in Manchester in 1798, unable to speak a word of English.

Nathan was to prove himself the outstanding financial genius of the family, and his leaving Frankfurt was to be the precedent for three of his brothers. Amschel remained in the old city; Solomon established a branch of the house in Vienna, Carl Meyer another in Naples, and James (Jacob) yet another in Paris. The international character of the great house was thus an established fact in the early decades of the nineteenth century.[5]

In 1803 William of Hesse was at last admitted to the Imperial College as Elector William I, and in the same year Meyer Amschel, with the assistance of Nathan in London, was able to negotiate a loan for Denmark for twenty million francs. This loan furnished both William and the Rothschilds with handsome profits, and similar transactions were made to the advantage of all parties during the ensuing nine years.

[5] The Naples house closed in 1861, after the creation of the Kingdom of Italy. The parent house, M. A. von Rothschild & Sons, at Frankfurt, closed in 1901, when the male line of the German branch of the family failed with the death of Baron W. C. von Rothschild, old Meyer Amschel's nephew. The Austrian house, once the most powerful bank in Europe, was not re-established after the Second World War. The French house, Messieurs de Rothschild Frères (Banque Rothschild), and the English, N. M. Rothschild & Sons, are, of course, still flourishing, though the former is now nationalized.

After the crushing defeat of the Prussians at Jena in 1806, the Elector fled to Denmark, and his kingdom was incorporated into the Confederation of the Rhine. William hastily confided a large part of his fortune, possibly £600,000, to the care of Meyer Amschel, who was able, despite threats of arrest and violence, and the offer of various bribes, to keep faith with his patron, and to restore his property, considerably augmented through careful investment, upon his restoration to the throne in 1813.[6] In the same year Meyer Amschel died, and from that time the Elector entrusted all his major financial business to the brilliant Nathan, who had become a naturalized British subject in 1804.

It will be seen from this account that well before 1815 the Rothschilds, and in particular Nathan, had become the principal loan contractors for a number of European governments. Efficiency coupled with absolute integrity—a hallmark of the house at every stage of its progress—had placed the family at the head of their profession. During the period of the Congresses, they made further very profitable transactions with most of the major governments of Europe, ultimately securing the confidence of the arch-conservative policeman of Europe, Metternich. Many honours, signs of approbation, and acknowledgements of their very real power were bestowed upon them.[7] Their rise to eminence, the essentially romantic character of their story, and the growing public awareness of their influence made it perhaps inevitable that they should come to the notice of the novelists of the day.[8] More

[6] The Elector's money was skilfully invested, with his full knowledge, by Nathan. Later legends romanticized these transactions, presenting Meyer Amschel as the faithful retainer, returning William's chests of plate intact after his restoration. See, for instance, the stylized miniatures of this subject reproduced to face p. 32 of *Cowles*. The details of the devious and complex negotiations are admirably set out in *Corti*, Ch. 3. During the period of the Napoleonic Confederation, Meyer Amschel was able to maintain good relations with the French authorities, and particularly with Prince Dalberg, the Primate. But he worked secretly at all times to preserve Elector William's fortune.

[7] In September 1816 Emperor Francis granted Amschel and Solomon the rank of primary nobility ('von'); Carl and James received the same honour in October. Branches of the family outside Germany used the French equivalent 'de' to indicate this same honour. Nathan became Austrian consul in London in March 1820; James became Consul-General at Paris in August 1821. On 29 September 1822 all five brothers were raised to the rank of hereditary baron of the Empire. Nathan, a British subject, preferred to be plain 'Mr', though his son Lionel used this Austrian title.

[8] Among the poets Byron links Rothschild and Baring in *Don Juan* (Canto XII, verse 5).

> Who holds the balance of the world? Who reign
> O'er congress, whether royalist or liberal? . . .
> Jew Rothschild, and his fellow-Christian, Baring.

These lines were composed *c.* 1822, four years after the two firms had attended the Congress of Aix-la-Chapelle.

important, however, is the fact that the most influential and powerful of the sons, Nathan, was a British subject. As the leading luminary of the City, the first financier to introduce foreign Consols into the London market, he thus created the new, international, character of City finance virtually single-handed.

To the historian of business, Nathan is a brilliant financier, constantly ahead of his time, and exhibiting an adventurous daring that at no time descended to mere speculation. To the novelist he was a disturbing phenomenon, frequently to be equated with the despised 'moneylender' of literary tradition, but at the same time looked upon with awe, and not a little fear. Furthermore, he was a foreigner, and a Jew;[9] despite his manifest patriotism, and his decidedly unromantic exterior, he had those characteristics—power, novelty, romance, awesomeness, and Semitic foreignness—that made him an ideal basis for literary depictions. In addition, in an age given to moralizing, he could provide a fund of metaphors for the castigation of the worship of Mammon.

Several novels appeared during the century containing characters either modelled directly on members of the Rothschild family, or motivated for part of their theme by the ideas attaching to that name.[10] In 1843 Mrs Gore issued her novel *The Moneylender*.[11] A year later Disraeli's celebrated *Coningsby* appeared. Both novels reflect a contemporary preoccupation with the Rothschilds and with the fascination of high finance.

The Moneylender concerns the friendship that develops between Basil Annesley, an honourable young man of aristocratic family, and Abednego Osalez, a moneylender of Jewish extraction. The Jew appears to Basil to be vengeful and harsh, delighting in the distress of his clients. We learn later that he is embittered both from suffering youthful humiliations in Parliament and from the landed interest, and from disappointment in love. He is the scion of a prominent Spanish Jewish merchant family settled in Cadiz, and is fabulously rich. In London he has many secret retreats, where he conducts business as a

[9] The prevailing views on 'Mammon' and the Jews are discussed in E. Rosenberg, *From Shylock to Svengali* (Stanford, 1961), notably Ch. 11; 'What News on the Rialto?'

[10] The first novelist of note to create a reflection of the Rothschilds was Honoré de Balzac, whose Baron de Nucingen is based upon James (Jacob) Rothschild of Paris, banker to Louis XVIII and Louis-Philippe, and the acknowledged king of high finance in Paris. Nucingen appears in *Père Goriot* (1835), *César Birotteau* (1837), *La Maison de Nucingen* (1838), and other works.

[11] Page references to *The Moneylender* are to the first edition, 3 vols. London, 1843.

rather squalid, shabby moneylender; but he also has a splendid town house,[12] and premises in the City, from which he conducts his 'respectable' business of stockbroker under his own name, rather than the sinister initials A. O., which he uses in his guise of moneylender.

We learn too that the moneylender is a Jew in race only: by religion he is a Christian, and his moneylending pursuits are in part a means of showing the more shallow members of the aristocracy the error of their ways by confronting them with their profligacy when they come to borrow further sums. He becomes more and more softened under Basil's influence, especially as Basil's mother is the woman who originally rejected Abednego through family influence. The plot has many Gothick convolutions and far-fetched coincidences—as Mrs Gore all but promises in her introduction. There is ultimate reconciliation between Abednego and those whom he has injured, and the novel has some telling portraits of contemporary aristocratic life, with its social pretences, and at times desperate nearness to bankruptcy.

Coningsby needs no summary here. However, in this novel, and in his *Tancred* which appeared in 1847, Disraeli introduces his famous Sidonia, perhaps the best-known reflection of a Rothschild in English literature. A comparison of these two fictional Jewish financiers will show that their creation could well have been stimulated by the life and career of Nathan Meyer Rothschild.

It is interesting to note that both Disraeli and Mrs Gore ascribe Sephardic origin to their financiers. The Jews of the Dispersion fall into two groups or ranks: the Sephardim, whose names are predominantly Spanish, Portuguese, and Italian; and the Ashkenazim, who are characterized mainly by German, Polish, and Russian names. Social superiority has traditionally attached to the Sephardim, who are granted a status akin to aristocracy. It is inevitable that Disraeli's

[12] As always in such novels, the houses of the business men have symbolic importance. Mrs Gore, who equates the West End with frivolity and lack of substance, is careful to place Osalez's opulent mansion in Bernard Street, Russell Square, which at that time was a little 'out of the way'. The rich appointments are described, and the food served at dinner is of a perfection 'unknown . . . on the Western side of Temple Bar!' (ii, p. 163.) Disraeli's financier Sidonia is not seen at home in London, but he has a splendid house in Paris, in the Faubourg St Germain, which had formerly belonged to the Grillon family. Osalez, too, has 'a princely hotel in the capital,—a noble country residence, once royal, situated on the wooded shores of the Seine' (iii, 3, p. 85). At the time these novels were written, Lionel de Rothschild was living at 107 Piccadilly. James de Rothschild lived in the Rue Saint-Florentin, in the former town house of Talleyrand; twenty-five miles east of the capital lay his palatial country seat of Ferrières.

Sidonia should be Sephardic, as his creator was immensely proud of his own Italian Sephardic origin.[13]

Mrs Gore, in her earlier novel, conforms to the same social convention, so that the reader of both novels is confronted at the outset with the aristocratic ring of Spanish Sephardic names, Sidonia and Osalez. Osalez is the only son of 'a wealthy merchant of Cadiz, trading with the whole commercial world, but chiefly with England.' (*The Moneylender* iii, Ch. 1, p. 34.) It will be important to the development of her novel that Osalez should not be Jewish at all, and so we learn that his grandfather had married, in England, 'a young Protestant of honourable extraction' (p. 34). His father had effected a similar alliance, so that Osalez was, in reality, more an English Protestant than a Spanish Jew.[14] Nevertheless, the Sephardic connection bestows upon the financier an aristocratic cachet that he shares with Sidonia and that could not be claimed by the real-life Nathan Meyer Rothschild and his brothers.[15]

Mrs Gore is content to make the founder of the Osalez family a merchant, albeit 'rich as a Doria or a Medicis.' (iii, Ch. 1. p. 35.) Disraeli, however, indulges in a romantic fantasy when he depicts Sidonia as a scion of an ancient, influential, and historical aristocratic family. Thus whereas the Osalez family encounter prejudice and danger from the 'illiterate bigots' of Cadiz, Sidonia, at least in *Coningsby*, is deliberately removed from any suggestion of the parvenu, who must struggle against the prejudices and exclusiveness of the landed interest.

Sidonia is descended from 'a very ancient and noble family of Aragon', one of a caste derived from the 'mosaic Arabs' who came to

[13] Disraeli always claimed that his family were originally Spanish Sephardim, expelled in the great persecution of 1492; his claim has been decisively disproved. (See R. Blake, *Disraeli* (1966), Ch. 1.)

[14] 'Basil noticed that, on a low table beside the flock bed, lay . . . a large crucifix of Berlin iron . . . A CRUCIFIX! The world, then, and his own suspicions, had decided wrongfully! Abednego the Moneylender was only in name and practice a Jew!' (ii, 2, p. 77.) One is reminded of the revelation in Maria Edgeworth's *Harrington* (1817) that Berenice Montenero (another Sephardi) is in reality 'An *English* Protestant!' (See also Rosenberg, op. cit. p. 66.)

[15] It will be appreciated that the Rothschilds were Ashkenazim, German Jews, who, though long established in Frankfurt, were of rather humble and obscure origin. Although *DNB* says that several members of the family were 'distinguished rabbis' in the seventeenth and eighteenth centuries, it should be remembered that very little is really known about their origin. Early accounts abound in inaccuracies, and in elaborations encouraged by Meyer Amschel himself after the family was ennobled.

Spain centuries before the Arabs of Islam. Secretly adhering to the Jewish religion while openly professing Christianity, they held many high offices in the kingdom. Disraeli chronicles their history in Book Four, Chapter Ten of *Coningsby*, where he further insists on Sidonia's noble origin by linking his ancestors with the Spanish ducal house of Medina Sidonia.[16]

Both authors furnish us with an account of their financier's education. Sidonia's father had emigrated to England, and Sidonia had thus been reared as an English boy. 'An Englishman, and taught from his cradle to be proud of being an Englishman, he first evinced in speaking his native language those remarkable powers of expression . . . which ever afterwards distinguished him.' (*Coningsby*, iv, Ch. 10.)

Mrs Gore's Osalez was already more than half English, and a Christian, so that there was no difficulty over entry to Eton and Oxford. At both school and university Osalez showed the same quality of intellectual brilliance that was to characterize the later creation, Sidonia.

At Eton 'I became an object of general wonder; my English verses, my proficiency as a Grecian, being equally themes of praise' (iii, Ch. 1. p. 43.) At Oxford too the future financier shone intellectually: 'The University had fixed its *imprimatur* on my scholarship . . .' (iii, Ch. 2. p. 57).

Sidonia, as a Jew, was barred from public school and university, and Disraeli provides him with a private tutor, Rebello. 'A Jesuit before the revolution, since then an exiled Liberal leader, now a member of the Spanish Cortes, Rebello was always a Jew.' (*Coningsby*, iv, Ch. 10.) This mysterious Jewish Liberal is yet another example of Disraeli's preternatural 'Arabs', a far cry from the Eton masters of Abednego Osalez.

Mrs Gore vividly describes what must have been a common experience for a Christianized Jewish youth—the sneering and arrogant persecution at both school and university arising from racial origin and characteristics. Such experiences play an important part in perverting the altruistic ideals of the young Osalez. The gilded Sidonia is spared any such vulgar humiliation.

[16] Disraeli's romantic account is, of course, based upon fact. A convenient account of the 'Nuevos Christianos', and the role of the Spanish Inquisition in their persecution, will be found in A. S. Turberville, *The Spanish Inquisition* (1932), particularly Ch. 2. It was the seventh Duke of Medina Sidonia (1550–1615) who so disastrously commanded the Spanish Armada in 1588.

Sidonia proves to be as brilliant intellectually as Osalez.

> The young Sidonia penetrated the highest mysteries of mathematics with a facility almost instinctive . . . The circumstances of his position, too, had early contributed to give him an unusual command over the modern languages. (iv, Ch. 10.)[17]

His further education was infinitely more exotic than that of Osalez at Oxford. For a period of five years he left his vast business concerns in the hands of others, and travelled to every corner of the globe, emerging from this grand tour a complete polymath (iv, Ch. 10). Osalez never attains such giddy heights, as Mrs Gore is always careful to give her fictional creations a firm basis in reality. However, to the historian of business it is disappointing to find that Sidonia, for all his preternatural brilliance, appears to do so little of financial interest if, indeed, he does anything at all.

It is in the early business history of the Sidonia family that the reader finds clear analogies to the Rothschilds and their rise to eminence. A sudden jump in historical time brings the reader of *Coningsby* from the age of the Inquisition to the era of the Peninsular Wars. In Book IV, Chapter Ten, we learn that during these wars 'a cadet of a younger branch of this family made a large fortune by military contracts and supplying the commissariat of the different armies.'

It is well known that Nathan Meyer Rothschild was largely instrumental in relieving the great financial embarrassment experienced by Wellington in the Peninsula. As early as 1809 Wellington had warned the government that lack of funds would oblige him to withdraw,[18] and by 1811 he was ruinously involved with questionable bankers in Malta, Sicily, and Spain. Bullion ships had been dispatched from England, which, if they passed Napoleon's blockade, sometimes sank in the Bay of Biscay. Nathan solved this problem in a daring and original manner. He bought up large quantities of Wellington's unpopular bills, cashed them at the Treasury, and sent the guineas thus acquired directly to France, where one of his brothers, usually James, paid them into several different Paris banking houses. Thence, through an intricate network of firms,

[17] Knowledge of modern rather than classical languages was essential in commerce, though evidently rather a novelty elsewhere. Mrs Gore describes Osalez's business guests as having 'a familiarity with modern languages. French, Italian, German, were familiar to them as English . . .' (*The Moneylender*, ii, Ch. 5, p. 176).

[18] *Wellington Dispatches*, iv, p. 473. (Letter to Huskisson, 28 July 1809.)

mostly Jewish, good bills quickly reached Wellington in Spain, where he converted them into gold.[19]

The British government, realizing the exceptional qualities of this young merchant banker, began to negotiate directly with him. Securing a large purchase of bullion from the East India Company, Nathan sold it to the Government and then, by further complex and devious routes, conveyed the bullion to Wellington in Spain.[20] Using his network of European connections, Rothschild was on other occasions able to transmit payments to other allied forces, for instance the Hessian mercenaries in the service of Austria.[21]

These transactions of Nathan's thus closely parallel the activities of the elder Sidonia as described by Disraeli in *Coningsby*. The next stage in the fortunes of the Sidonia family is the emigration to England of Sidonia's father. 'At the peace . . . Sidonia . . . resolved to emigrate to England, with which he had, in the course of years, formed considerable connections . . . He arrived here after the Peace of Paris [1815] with his large capital.' (IV, Ch. 10.) Nathan had emigrated to Manchester in 1798, moved to London in 1803, and established his reputation securely as one of the foremost bankers of the capital twelve years before the supposed arrival of the fictional Sidonia. However, the broad similarities between the movements and careers of the real and the fictitious banker strongly suggest that Disraeli was thinking of Nathan Rothschild when he described the doings of the elder Sidonia.

We are told that Sidonia, once in England, 'staked all that he was worth on the Waterloo loan, and the event made him one of the greatest capitalists in Europe.' (iv, Ch. 10.) This statement is clearly a reflection of the legend that Nathan Meyer Rothschild founded his fortunes on enormous loans made to the Government during the Napoleonic crisis. When on 14 June 1815 the Chancellor of the Exchequer, Vansittart, brought in his war-footing budget, he was compelled to raise an additional £39 million through loans. This presumably, is the

[19] *Corti*, pp. 115–17. The French government was perfectly happy to allow English gold to enter France, as it was seen as a proof of England's progressive decay. It never seemed to occur to it that gold securely lodged at Paris could be used to guarantee mobile paper. See in particular Marion, *Histoire Financière de la France depuis 1715* (Paris, 1914), iv, pp. 358 ff.

[20] *Corti*, pp. 117–18. For an antagonistic account of this transaction, see Fox Bourne, *The Romance of Trade*, Ch. 11.

[21] *Corti*, p. 125.

'Waterloo loan' upon which the elder Sidonia founded his spectacular fortune.[22] However, the greater part of Nathan's fortune, and indeed of the fortunes of all the Rothschild brothers, had been amassed by patient and complex negotiation of loans and transfers to the allied powers. The sudden, mad venture of the gambler 'staking his all' was alien to the thinking of the Rothschild family.

Fox Bourne's account of Nathan at Waterloo is typical of this persistent and erroneous fiction. Usually an astute historian of mercantile history,[23] he nevertheless recounts the Waterloo fiction:

> Rothschild was at Hougoumont on the memorable day of the battle of Waterloo, watching its incidents . . . Immediately the issue of the battle was clear, he hurried off, and, with help of carriages and horses . . . reached England a day before any other messenger arrived. During that day his secret agents, on the strength of a report, said to have originated with him, that the English forces were defeated, bought at panic prices, a cartload of shares, bonds, stock and scrip . . . which, as soon as the true news arrived . . . he was able to sell again for their real value; thereby, it was rumoured, he cleared £1,000,000.[24]

This account could well apply to the elder Sidonia, who 'staked all' over Waterloo; it is in every detail untrue of Nathan. He was in England at the time of the battle, in consultation with Herries, the Commissary-General. Although he obtained news of the English victory very early on the morning of 20 June through his superior private courier system, the Government refused to believe him, and waited until Wellington's official envoy arrived during the course of the next day. The story of the 'cartload of shares' and the £1 million is an apocryphal transaction, which is known to be totally without foundation.

Both Government and public were enormously impressed by Nathan's advance knowledge of the outcome, and this, linked with a growing public awareness of the importance of the Rothschilds to the Treasury throughout the wars, gave rise to the various legends, including that recounted as fact by Fox Bourne.[25]

[22] See J. F. Rees, *A Short Fiscal and Financial History of England (1815–1918)*, 1921, pp. 35–6.

[23] For instance, he dismisses the idea that the Rothschild fortunes were founded entirely on war-loans after 1810: 'Rothschild was also a contractor for the English Government; but his prosperity resulted chiefly from his introduction of continental loans into the English market. The vast trade in foreign Consols . . . owes its origin entirely to him.' (*The Romance of Trade*, p. 323.)

[24] Fox Bourne, pp. 323–4.

[25] My account of the Waterloo legend is derived from *Corti*, pp. 158–9.

The elder Sidonia's post-war financial career clearly parallels that of Nathan Meyer Rothschild. Realizing that an exhausted Europe 'must require capital to carry on peace', Sidonia made such capital available. 'France wanted some; Austria more; Prussia a little; Russia a few millions. Sidonia could furnish them all.' (iv, Ch. 10.)

It was in 1818 that the Prussian Chancellor, Prince Hardenberg, recommended that a state loan be raised in England through the agency of Nathan. The latter was able to oblige, and a loan of £5 million was secured. This was the first state loan raised by Nathan, and it was to be the first of many. Offered at 72, it was quickly taken up, and the prestige of Nathan was such that it never fell below this issue price, and in 1824 it actually reached par. This is a typical example of the kind of loan negotiated by the House in the post-war period. Certainly the brothers had constant financial dealings with the chancellors of the countries 'furnished' by Sidonia.[26]

From what has been written it may be seen that the elder Sidonia, as a financier, is a composite of the celebrated Rothschild brothers, with an emphatic echo in particular of Nathan. His origin as a cadet of a noble Spanish family, and as a Sephardi, one of the aristocrats of the diaspora, is to furnish him, or rather his son, with the kind of impeccable pedigree so beloved of Disraeli, and so necessary to the identity of his heroes and seers.[27] But in the world of European finance we are firmly in the ambience of the Ashkenazim, the socially-suspect German-Jewish family from Frankfurt. 'The son of a Spaniard' he may be; but Sidonia visits 'his uncle at Naples', and 'another of his father's relatives at Frankfort'. Frankfurt, London, Paris, Naples—these were the four bases of the German-Jewish House of Rothschild.[28]

Very similar connections between the House of Rothschild and Osalez are found in Book III, chapter 8 of *The Moneylender*. 'Osalez

[26] See in particular *Corti*, Ch. 4: 'The Brothers Rothschild During the Period of Congresses, 1818–1822'.

[27] '. . . why will Mr. D'Israeli be so fond of dukes?' wrote R. M. Milnes (*Edinburgh Review*, No. 86 (July 1847), p. 142). For a cogent explanation of Disraeli's views on aristocracy, see Blake, op. cit. pp. 278–84.

[28] At his death, the elder Sidonia 'was lord and master of the money-market of the world, and of course virtually lord and master of everything else. He literally held the revenues of Southern Italy in pawn . . .' (*Coningsby* iv, Ch. 10). Nathan was certainly 'lord and master of the money-market of the world' by 1818. His brother Solomon may be said with truth to have held the revenues of Naples 'in pawn' in 1821. (See *Corti*, p. 233.)

had left London in the Antwerp steamer. His letters were to be addressed to Frankfort on the Main . . . it was supposed that his ultimate destination might be Vienna . . .'.

Antwerp, Frankfurt, and Vienna, in their different ways, were all prime centres of Rothschild activity. In the same chapter we learn that 'the Hessian Government' are happy to grant a favour to 'the potent and influential Osalez, whose name was the support of their loans, and whose word was as that of a sovereign prince!' Such a direct reflection of the intimate relationship of the Rothschilds with the House of Hesse would seem to be conclusive evidence of Osalez's origin. However, realizing perhaps that too close a comparison would be invidious, Mrs Gore in the same chapter arranges a meeting between the reality and the reflection:

> [Osalez was seen] dining at Monrepos with the Grand Duke of Nassau,—at Amelienbourg with the Grand Duke of Baden,—and, more ennobling than all,—at Frankfort with the Grand Duke of Lucreland—Rothschild the Great.

Although the rather confusing chronology of the novel would suggest, together with their place of meeting, that Mrs Gore means this to be Meyer Amschel, it is certainly more likely that 'the Grand Duke of Lucreland—Rothschild the Great' is a hyperbole suggested by the native figure, Nathan Meyer Rothschild.

It is very interesting to observe how both Mrs Gore and Disraeli provide an authentic business ambience for their financiers. There is no description of Sidonia's place of business in *Coningsby*, but in the later novel *Tancred* (1847)[29] his premises are fairly closely located and detailed, and once again we can detect clear affinities with the London headquarters of Nathan Meyer Rothschild.

> In a long, dark, narrow, crooked street, which is still [1847] called a lane, and which runs from the south side of the street of the Lombards towards the river, there is one of those old houses of a century past, and which, both in its original design and present condition, is a noble specimen of its order. (p. 114.)

The 'long, dark, narrow, crooked street' is readily located as one of the group of ancient lanes, now severed diagonally by King William Street, that lie to the south of Lombard Street, and east of the Mansion House. They are clearly visible in their medieval huddle in Frans

[29] Page references to *Tancred* are to the edition of that novel in the Hughenden Edition of the *Novels and Tales* (1881).

Hogenberg's map of London (1572), and although totally destroyed in the Great Fire of 1666, the new buildings were raised on the same street plan. It was here, in St Swithin's Lane, that Nathan Meyer Rothschild established his London business in 1804.

Sidonia's house is 'of a century past', and we learn that it is called Sequin Court (p. 115).[30] Disraeli's description of the house has definite similarities to Rothschild's house, New Court, in St Swithin's Lane.

A pair of massy iron gates, of elaborate workmanship, separate the street from its spacious and airy courtyard, which is formed on either side by a wing of the mansion, itself a building of deep red brick, with a pediment, and pilasters, and copings of stone. A flight of steps leads to the lofty and central doorway; in the middle of the court, there is a garden plot, inclosing a fountain, and a fine plane tree. (p. 115.)

In similar fashion Rothschild's New Court, at least in its early days, was divided from its neighbours in St Swithin's Lane by railings, though without the 'massy iron gates'. New Court was arranged on three sides of its courtyard, but its original brick front seems to have been surfaced either with stucco or Bath stone. The main wing, facing up the lane, had its 'lofty door' above a flight of stone steps.[31]

Disraeli's description is designed to evoke a picture of a gracious, opulent mansion. New Court was, in fact, a solid, unpretentious block of buildings, combining Nathan's living quarters with his counting-house and bank.[32] In 1826 Nathan and his family moved to a spacious house at 107 Piccadilly, although the business of banking continued at New Court. Although altered and improved later in the century, the basic plan of a winged mansion with a courtyard persisted, and Disraeli, a life-long friend of Nathan's son Baron Lionel de

[30] The sequin was an Italian gold coin, minted originally at Venice in the thirteenth century. Last issued at Rome in 1834, the name as applied here adds a further 'romantic' dimension to Disraeli's exotic banker.

[31] A painting of New Court as it appeared in 1819 is reproduced on p. 52 of *Cowles*. The railings must have been removed at an early date. For confirmation of the general appearance of New Court in the early nineteenth century, I am very grateful to Mr C. D. H. Robson of N. M. Rothschild & Sons Ltd.

[32] Prince Puckler-Muskau visited Nathan at New Court in 1826, and was clearly not impressed with the building. 'I found him in a poor, obscure-looking place . . . and making my way with some difficulty through the little courtyard, blocked up by a waggon laden with bars of silver, I was introduced into the presence of the Grand Ally of the Holy Alliance.' (Prince Puckler-Muskau, *Tour of a German Prince* (London, 1832), iii, pp. 62–4.)

Rothschild, whom he had met in 1838, would have been familiar with the 'nerve centre' of the famous merchant bankers at New Court.[33]

Osalez the stockbroker occupied premises leading off Old Jewry, on the other side of the Mansion House from St Swithin's Lane, and thus firmly in the centre of the City: a short walk would have brought him to New Court or to 'Sequin Court'.

> Basil proceeded through the gorge of a narrow court into a larger one . . . one side of which seemed occupied by a handsome old-fashioned dwelling-house, and the other by a range of buildings, the basement story of which was appropriated to counting-houses. (ii, Ch. 7, p. 283.)
> . . . the paved open space adjoining the old mansion house, and ruralised by the name of garden, because containing a pump, and an old sycamore. (ii, Ch. 7, p. 287.)

It is interesting to see, in this fictional description of the premises of a great financier, the same lack of ostentation, amounting almost to obscurity, that characterized those of Sidonia and Rothschild. Again, there is the narrow court, the old-fashioned dwelling-house, the paved open space, the old tree. We may imagine that there were many such cramped quarters, occupied by honest and successful businesses, the names of which were a greater guarantee of their integrity than a palatial office building.

Both Mrs Gore and Disraeli are careful to emphasize the almost total lack of contact that prevailed between the energetic City life and the aristocratic dwellers in the West End. Both are aware of the enormous power being generated in the City, and of the growing dependence of Government upon the great financiers during and after the French wars. Each is aware too of the conflict growing between these 'devotees of Mammon' and the ancient landed interest. Disraeli looks for a spiritual renaissance in the aristocracy, and creates Sidonia partly as a reconciling link between Land and Capital, aristocrat and entrepreneur. Mrs Gore is basically a realist. While constantly denouncing 'Mammon'—the worship of money—she sensibly points out that capital is becoming the basis of power in contemporary society, and, while recognizing that many aristocrats are noble, she is suspicious of mean jealousies and snobbishness. Her 'hero', Basil Annesley, shrinks in stature as he leaves his patrician world west of Temple Bar

[33] The premises survived the blitz of 1940–1, standing alone amid the rubble of St Swithin's Lane. New Court was demolished in 1962, and the construction of the present glass and aluminium New Court was completed in 1965.

and ventures into the City. He inquires at Osalez's house in Bernard Street the whereabouts of the financier's place of business.

'And where am I to find him in the city?' demanded young Annesley; a query that appeared to excite as much amazement in the rotund pantler, as though he had demanded in what quarter of the town he was to look for Westminster Abbey. 'Mr. Osalez, Sir, will be on the Stock Exchange', said he, conceiving that the handsome young gentleman . . . must be infirm of intellect. 'If off 'Change, you will find him at his house of business.' 'And where is that?' incautiously inquired Basil. The man seemed to draw largely upon the decorum of his calling, in order to refrain from a laugh. 'In the Old Jewry, Sir.—But you need only mention the name of Mr. Osalez in the city, Sir, for anyone to shew you the way. The first cabman or orange-boy you meet will inform you.' (*The Moneylender*, ii, Ch. 7, pp. 279–80.)

In similar fashion Tancred, the young Lord Montacute, having penetrated into that mysterious and unknown area, the City, finds that the name 'Sequin Court' is synonymous with Sidonia, just as New Court could mean only the Rothschild bank.

'Do you happen to know, sir, a place called Sequin Court?'
'I should think I did,' said the man, smiling. 'So you are going to Sidonia's?' (*Tancred*, p. 118.)

In both extracts one notices the tolerant and humorous condescension of those connected with the pulse of the City towards a youthful, aristocratic intruder, whose ignorance of the realities of power is to be pitied. For Tancred a realization of the true worth attached to his title comes when he tells a porter, 'I want Monsieur de Sidonia', and receives the blunt answer: 'Can't see him now; he is engaged.' He has ventured into a different world, where values operate on another scale, and his reactions are identical to those of Mrs Gore's earlier creation: 'Basil Annesley found himself among a race of persons with whom he had neither an emotion nor an impulse in common . . .' (*The Moneylender*, ii, Ch. 7, p. 285.)

Although in all three novels there are many references to the great financial operations of Sidonia and Osalez, the reader finds little significant business transacted. In *Tancred* occurs the sole instance of Sidonia occupying himself with financial matters. To a confidential clerk he says: 'Write . . . that my letters are twelve hours later than the despatches, and that the city continued quite tranquil. Let the extract from the Berlin letter be left at the same time at the Treasury.

The latest bulletin?' The clerk replies: 'Consols drooping at half-past two; all the foreign funds lower; shares very active.' (*Tancred*, p. 124.)

This brief dialogue punctuates Sidonia's revelation to Tancred of the Great Asian Mystery; it functions principally to remind the reader—in case he has forgotten—that Sidonia is a man of business. The clerk's information is not particularly exciting: it is a routine afternoon report on the state of the stockmarket. There is, though, a little touch of Rothschild in the implication that the Treasury has come to expect Sidonia to furnish it with late news; in this case, presumably, the political temperature of Berlin as interpreted from the state of the market there.

More interesting is the episode in *Tancred* treating of the Baroni family.[34] Baroni, a travelling performer with a talented family, is patronized by Sidonia, and is appointed superintendent of the couriers.

> He himself was to repair . . . at once to Vienna, where he was to be installed into a post of great responsibility and emolument. He was made superintendent of couriers of the house of Sidonia in that capital, and especially of those that conveyed treasure. (*Tancred*, p. 335.)

This clearly echoes the famous system of couriers that Nathan had set up after his establishment in London in 1804. The network spread from London to the channel ports, thence to Ostend, Ghent, Brussels, and Amsterdam, and its great swiftness and efficiency caused the British government to make frequent use of it in preference to its own official system.[35]

None of these pieces of business, however, actually shows us Osalez and Sidonia swaying the fortunes of Europe. It is more important for these novelists to convey a single abstract idea—that of 'Mammon',

[34] Mr Richard A. Levine sees the Baroni family as the successful blending of East and West, a microcosm of the 'Hebraeo-Christian concept' that lies at the heart of the 'Great Asian Mystery'. 'The very blending which lay at the heart of the Asian mystery is seen in this family—the apostolical succession in terms of ethnic, national and religious commingling.' (R. A. Levine, 'Disraeli's *Tancred* and "The Great Asian Mystery"' in *Nineteenth Century Fiction*, No. 22 (1968), pp. 71–85.)

[35] See in particular *Cowles*, pp. 47–8, and p. 71. Nathan's appreciation of the value of early news was inherited from his father, Meyer Amschel. In the closing years of the eighteenth century he had established business relations with Karl Anselm Prince von Thurn und Taxis, Postmaster of the Holy Roman Empire. The Thurn und Taxis family had run the central European postal system since 1516. Letters were regularly opened, and important news transmitted to the Roman-German Emperor, and it was advantageous to be on intimate terms with the Postmaster, who was willing to share news derived from 'secret manipulation'. The headquarters of the Imperial postal service was at Frankfurt—a matter of great convenience for Meyer Amschel (*Corti*, pp. 29–30).

'Finance', 'Speculation'—and its concomitant, Power. They present the power of money as an object inspiring awe, and not a little fear. Sidonia had openly admired the power of Acquaviva, who 'ruled every cabinet in Europe', and colonized America before he was thirty-seven: 'the secret sway of Europe! That was indeed a position! (*Coningsby*, iii, Ch. 1.)[36] In the early, bitter, phase of his career, Osalez had equated power with gold alone, and expressed his belief in the following awesome words: 'MAMMON holds the preponderating influence . . . GOLD, GOLD, GOLD, constitutes . . . the sole divinity—the Jehovah of the universal earth!' (*The Moneylender*, iii, Ch. 2, p. 78.)

It is in such dramatic hyperbole, rather than in the delineation of a typical international transaction, that these novelists convey their intentions. Mrs Gore can be particularly melodramatic. Observing the 'puissant old men' assembled for dinner at Osalez's house in Bernard Street, Basil imagines to himself how the financier would describe his guests:

> Behold! these be the kings of whom I spake! . . . These are the master-hands that move the wires of kingly puppets,—these are the mainsprings of aristocratic action,—these are they without whom privy-councils and parliaments might mouth and gibber in vain,—these are the veritable monarchs who make peace and war . . . (ii, Ch. 4, p. 168.)

Another character declares that the very name of Osalez can 'convulse the Stock Exchanges of the various capitals of Europe' (ii, p. 230).

Sidonia and Osalez as financiers share a particular kind of common origin. Neither is a direct portrait of Nathan Meyer Rothschild, but the foregoing discussion has shown that the career of this great international financier had motivated their creation. Various attempts have been made to seek in real life for a specific analogue to Sidonia, with varying degrees of feasibility.[37] An attractive possibility is

[36] Claudius Acquaviva (1543–1615) was fifth General of the Jesuits. Both as organizer and educationalist, he was the most able General the Order ever had, having been elected at the early age of 37. His educational influence is still felt in the Jesuit Order, though Sidonia's—and Disraeli's—claims about him are grossly exaggerated.

[37] The revised key to *Coningsby*, issued in 1845, states that Sidonia is modelled on 'Baron Alfred de Rothschild of Naples', and Professor Asa Briggs accepts this in his edition of the novel (Signet Classics, New York, 1962, p. xiii). I can find no member of the Rothschild family bearing the name 'Alfred' prior to Nathan's grandson Alfred, born in 1842. The head of the Neapolitan house in 1844, when *Coningsby* appeared, was Baron Carl Meyer von Rothschild, then aged 56, sixteen years Disraeli's senior. Wilfred Meynell considered that Sidonia 'stood as a type of' the Rothschilds, and C. L. Cline saw him as 'Disraeli blessed with the Rothschild millions'. (W. Meynell,

Disraeli's life-long friend Baron Lionel de Rothschild who, on Nathan's death in 1836, had brilliantly assumed full control of the London house.[38] Refined and aristocratic, he was a marked contrast to his rotund and blunt father, Nathan. He successfully applied in 1836 for permission to use the Austrian title of baron, which had been bestowed upon his father in 1822, and in common with other Rothschilds of his generation he was a cultured patron of the arts.

Undoubtedly attractive to Disraeli was Lionel's devotion to his Jewish race and religion, a principal characteristic of Sidonia. He had co-operated in 1843 with the Jewish banker and philanthropist Sir Moses Montefiore in an attempt to lighten the lot of the Jews of Russia and Poland, something which Sidonia had also undertaken. He tells Coningsby of 'our representations in favour of the Polish Hebrews, a numerous race, but the most suffering and degraded of all the tribes, [which] have not been very agreeable to the Czar' (Bk. iv, Ch. 15).

These considerations incline one to see Sidonia as a combination of the thrusting power and legendary character of Nathan Meyer, with the youthful, refined, and philanthropic qualities of Disraeli's close friend, Lionel de Rothschild.

So far, Sidonia and Osalez have been examined solely as reflections of great contemporary financiers. However, they both transcend this limiting characterization. A particularly succinct sketch of Sidonia is given by Oliver Elton.

> We meet the preposterous, the all-accomplished Sidonia, with his callous heart and searching brain, with one hand on the purse-strings and the other on the policy of Europe; a Jew's heated daydream of the ideal Jew. But Sidonia is also the mouthpiece of the huge and bombastic, but not untrue, tirade upon the Hebrew genius, which is at once a confession of faith and a protest.[39]

Benjamin Disraeli, an Unconventional Biography (1903), i, pp. 21–2; and C. L. Cline, 'Disraeli and Thackeray', *Review of English Studies*, vol. 19, No. 76 (1943), p. 406.)

[38] During his career, Lionel arranged eighteen important government loans, including the Crimean loan of 1856, which drew £16 million. He assisted the American and Russian governments, and was a principal financier of the Austrian Empire. He financed the French and Italian railway systems, and in 1876 advanced the British Government the money necessary to purchase the Suez Canal shares. He is equally well remembered for his success in entering Parliament as member for the City in 1858, eleven years after his election, having been barred by the Lords from taking his seat because of his inability to take a Christian oath at the bar. He died in 1879.

[39] O. Elton, *A Survey of English Literature 1830–1880* (1927), ii, pp. 179–80.

Both Sidonia and Osalez are indeed 'all-accomplished', equally at home on Parnassus or in Lombard Street, and this amalgam of qualities causes each of them to be cast in the role of tutor to a young man of aristocratic background, totally ignorant of the ways of the financial world.

From the opening chapter of Book III of *Coningsby*, where he is revealed to us literally in a flash of lightning, Sidonia converses with the young nobleman in a tone characterized by calm authority, unexpected paradox, and philosophical didacticism. It would be inaccurate to speak of a 'conversation' between the two 'young men'[40]: in reality it is a tutorial, in which Sidonia is seen to be a don of quite exceptional talents, to whom Coningsby defers:

> On all subjects his mind seemed to be instructed and his opinions formed . . . he solved with a phrase some deep problem that men muse over for years. He said many things that were strange, yet they immediately appeared to be true. (iii, Ch. 1.)

When the 'conversation' ends, Coningsby instinctively confesses that he regards himself as Sidonia's pupil: 'But [my acquaintance] is not worth preserving . . . It is yours that is the treasure. You teach me things of which I have long mused.'

This teacher–pupil relationship is developed at length in the celebrated interview in Book IV, chapter 15 of *Coningsby*, where the young man is given a lecture on Jewish pre-eminence and racial superiority. Later, in *Tancred*, Sidonia sends his second pupil forth to carry out practical research into the 'Great Asian Mystery'. Sidonia is thus the instigator of action in others; he has no active role in the events of either *Coningsby* or *Tancred*. Action, as he himself tells us, is not for him.[41]

Osalez, on the other hand, is the prime mover of the whole story of *The Moneylender*. But he, too, fulfils the role of tutor to a young nobleman, Basil Annesley. Mrs Gore engineers a number of situations in which Osalez and Basil are alone, and where the older man can

[40] Although described by Disraeli as being 'perhaps ten years older than Coningsby', and so in his mid-thirties, Sidonia is none the less 'in the period of lusty youth' (*Coningsby*, iii, Ch. 1).

[41] 'His place in *Coningsby* is that of an adviser to the hero, not that of a mover of the plot. Only twice, on unimportant matters, does Coningsby act on Sidonia's advice, and the result is 'neither good nor bad for the hero'. A. H. Frietzsche, 'Action is Not for Me' in *Utah Academy of Science, Proc.* (1959–60), p. 48.

castigate the hypocrisy of an age which pretends that money is something to be despised. When Basil ill-advisedly speaks of 'mere wealth', the expression elicits from Osalez a lecture on the all-pervading power of money:

What but money causes the crucible to glow,—sinks the shaft,—launches the balloon into the sky—or plunges the diving-bell into the depths of the ocean?—Of what metal is composed the key of the poet's imagination—the orator's eloquence—the physician's skill—the divine's zeal and fervour?—Of gold, Sir—of current gold! etc. (i, Ch. 7, p. 262.)

The lesson is reinforced on later occasions, and Basil could not have complained that the realities of life were kept from him by this eloquent and effective scourge of cant:

I command most of those who command the destinies of the kingdom—and even as war-making kings tremble under the governance of Rothschild, under *mine* . . . abide more than one, two or three of those to whom you uncover your head reverentially as you pass! (ii, Ch. 1, p. 34.)[42]

A further similarity between Sidonia and Osalez is their looking to 'the East' as the fountain of wisdom. Sidonia, who had travelled extensively in Palestine, has clearly discovered for himself the interpretation of the 'Great Asian Mystery', and is equally certain that his pupil will do the same. Osalez, too, as a young man, travelled in the East, where he was welcomed by the highest ranks of 'Arabian' society. 'Those courts which we—the real barbarians—tax as barbaresque, are more refined in their indolent resources than our pompous Western show of sovereignty . . . No false jewels,—no gilding,—no tinsel . . .'. (iii, Ch. 2. p. 81.) Even nearer in sentiment to Sidonia's racial theories is the substance of the following passage:

I wanted to look upon the land which had given birth to my ill-fated race . . . in the East, I thoroughly threw off the prejudices of civilization . . . to behold other creeds established as firmly . . . convinced me that the all-seeing God . . . must behold with sentiments of mercy . . . the hereditary responsibility of the children of Israel for the predestined crime of their

[42] The inclusion of tutor–pupil relationships in these novels is a symptom of an age seeking for new gods, a synthesis of social ethics and the uncheckable march of industrialism. The most ludicrous of these mentors is surely Kingsley's Barnakill in *Yeast* (1851), who plies Lancelot with vapid platitudes, and in Ch. 17 spirits him away from St Paul's Cathedral to a mysterious land, or state, built upon 'Jesus Christ—THE MAN', presumably a vaguely-realized Christian Socialist Utopia.

forefathers! Thenceforward the Jews . . . became, in my eyes, as any other people; save in being more unjustly aspersed, and consequently more deserving commiseration . . . (op. cit., iii, Ch. 2.)[43]

Mrs Gore here pleads for the Jews to be regarded 'as any other people'; Disraeli, however, turns to the East and its ancient races for purposes which cause Sidonia and Osalez ultimately to part company. Sidonia is the means whereby Disraeli heralds the genius of the Hebrew people, and makes his plea for the nobility and value of a race still at that time despised, ridiculed, and vilified in a way that would be totally unacceptable today. Further, he is the vehicle for Disraeli's serious attempt to create a religious philosophy that would reconcile Judaism and Christianity. Such a reconciliation is symbolized in *Tancred* by the love and presumed later marriage of the young Lord Montacute and the 'Arabian maiden' Eva.[44] Their union, that of an aristocracy of birth with one of wealth and merit, should lead to a joint sense of responsibility for their 'inferiors'. But this state cannot develop without a new, non-political impetus, a spirituality unfettered by the arid tenets of Utilitarianism.[45] Tancred confesses to Eva that money is the ruling principle of his country; he had therefore to set out to discover a new creed of 'great principles' to invigorate English life, a creed based upon 'the amalgam of Hebraism and Christianity found in the Primitive Church'.[46]

[43] It is very tempting to suggest a direct link between this passage and Disraeli's similar material in the later novel, *Coningsby*. However, it is as well to remember that 'the East' was still a source of romantic interest at the time that both novels were written. Thomas Moore's *Lalla Rookh* (1817) was constantly being re-issued, and there were editions in 1834 and 1844. Lady Hester Stanhope had died in 1839, but her romantic life of self-exile on Mount Lebanon in Syria still lived in the public imagination. In 1844 Alexander Kinglake published his memorable description of her in *Eothen*, and in the following year her memoirs were published. Mrs Gore and Disraeli shared the same publisher, Henry Colburn of New Burlington Street. One cannot, though, assume that Disraeli knew Mrs Gore's work. He did not read much contemporary literature (Blake, *Disraeli*, (1966), p. 191). It would be prudent to see Mrs Gore's eastern material as forming part of a current interest.

[44] Similar symbols of reconciliation occur in *Coningsby*, where the marriage of Coningsby and Edith Milbank link aristocracy with trade, and in that of Egremont and Sybil in *Sybil* (1845), which reconciles New Toryism with populist aspirations.

[45] In *Tancred*, Lady Bertie's desire for a railroad to Jerusalem symbolizes the inability of the Utilitarian to achieve a spiritual goal.

[46] Quoted from Richard A. Levine, 'Disraeli's *Tancred* and "The Great Asian Mystery"', *Nineteenth Century Fiction*, No. 22 (1968), pp. 79–80. My views on the 'Great Asian Mystery' in this chapter owe much to Mr Levine's brilliant essay.

This idea of a Hebrew–Christian Church as an overwhelming power for spiritual regeneration in a materialistic age is a noble one, however naïve it is theologically and practically.[47] However, its tenets, and Disraeli's exaltation of the Jewish race as the means of salvation, were received in his own day with outrage and rebuke, and Sidonia, as the instructor of Coningsby in racial matters, and the direct instigator of Tancred's quest for the solution to the Great Asian Mystery, is remembered for these traits, and not principally for his connections with the Rothschilds.

Disraeli's attempt to plead the genius of the Jewish race met with some fierce responses. R. M. Milnes saw it as an obscurantist exercise in 'fanaticism and charlatanism',[48] and there is no doubt that Sidonia's racial harangues are as disturbing to the modern reader as they were to Disraeli's contemporaries.[49] Milnes further saw the novelist's ideas as an attack upon nationalism and patriotism; a slandering of the British people as 'jaded slaves', whose civilized progress and self-reliance are 'here proclaimed an entire delusion and failure'. Disraeli's Hebraic-Christian Church 'neutralises the last eighteen hundred years of the world . . . [It] would substitute a quietist adoption of absolute *a priori* impressions for the fruits of the laborious analysis of generations.'

Milnes is clearly disturbed at the damage that these ideas could inflict upon the English Jewish community. He states that the Jews, while believing that they have a peculiar destiny, are content to 'discharge the duties of citizens in a free state', and draws a pointed distinction between the prominent citizen Lionel de Rothschild and the self-appointed pariah, Sidonia: 'Baron Rothschild may become member for the City; but Sidonia, the ideal Jew, must remain an alien till he returns to Jerusalem.'

There were other, less moderate, responses to Sidonia and his ideas. G. H. Lewes, reviewing the fifth edition of *Coningsby* in 1849, reveals

[47] 'Disraeli was so impressed with the awesome mystery of the universe that he could not be satisfied by attempts to explain the mystery in finite terms, whether such attempts be made by metaphysicians or theologians.' (C. J. Lewis, 'Disraeli's Concept of Divine Order', *Jewish Social Studies*, No. 24 (1962), p. 145.)

[48] R. M. Milnes 'Tancred', *Edinburgh Review*, No. 86 (1847) pp. 138–155. The review was not signed.

[49] 'Disraeli harps on the racial supremacy of the Jews in a manner that nineteenth-century intellectuals found vapid and that twentieth-century Jews, with the experience of the Master Race behind them, find embarrassing . . .' (Rosenberg, *From Shylock to Svengali*, pp. 177–8).

that he had so completely rejected all Disraeli's philosophies and ideals that he had convinced himself that Sidonia's creator was a charlatan. He writes twenty pages of vituperation and quite incredible abuse, calling Disraeli a 'disreputable and disrespected person', who wrote 'verbiage—nothing else'. He is an 'adventurer' and 'charlatan', whose 'passion is mere sensuality', and whose only achievement is to create 'a white-waistcoat philosophy to adorn novels and historic fancies'. Lewes strongly condemns what he sees as a sinister racialism in Disraeli's work: 'Sidonia will demonstrate to you that the Jews are the greatest and grandest specimens of the human race, and by prescriptive right divine, must and will rule it.'[50] Such words from an important critic, author, and philosopher show a total lack of communication between Disraeli and his more thoughtful contemporaries, and reveal just how ill attuned to the times Disraeli had been in creating Sidonia, not as another Osalez, but as the promulgator of new and outrageous spiritual ideas.

Another reaction to Sidonia was satirical banter. Thackeray had written a very fair review of the first edition of *Coningsby* in the *Morning Chronicle* and, while castigating Disraeli's racial theories, had admired the novel sufficiently to desire a meeting with its author.[51] The two men seem to have met, but never on terms of positive friendship, and any possibility of such amity died after the appearance in *Punch* in 1847 of Thackeray's burlesque *Codlingsby. By B. de Shrewsbury, Esq.*[52] This story of the old-clothes-seller Rafael Mendoza (who is, of course, Sidonia) is by any standards not particularly successful, and the twentieth-century reader cannot be amused at the cheap anti-semitism that informs it. One part of the burlesque is, however, just, and delivers Disraeli a well-deserved rebuke.

It will be recalled how, in Book IV, chapter 15 of *Coningsby*, Sidonia lectures the young nobleman on Jewish pre-eminence: 'the Jewish mind exercises a vast influence on the affairs of Europe . . . The first Jesuits

[50] G. H. Lewes, *British Quarterly Review*, No. 10 (1849), pp. 118–38. Lewes's review of *Coningsby* echoes the contempt and utter disbelief of an earlier review of *Tancred* by J. Russell Lowell, *North American Review*, 65 (1847), p. 83: 'The work by which the elder D'Israeli will be remembered is the old curiosity shop of literature. He is merely a cast-off-clothes-dealer in an aesthetic sense. The son, with his trumpery of the past, is clearly a vender of the same wares, and an offshoot from the same stock.'

[51] For the relations between Disraeli and Thackeray, see in particular C. L. Cline's essay 'Disraeli and Thackeray', *Review of English Studies*, xv (1943), No. 76, pp. 404–8.

[52] Disraeli was at the time MP for Shrewsbury. In later editions the ascription was altered to read 'By D. Shrewsberry, Esq.'

were Jews; that mysterious Russian diplomacy which so alarms Western Europe is organised and principally carried on by Jews . . .' Sidonia then instances the European finance ministers who, he claims, are Jews. It is this celebrated 'list' that Thackeray eagerly seized upon in his burlesque. Codlingsby is astounded to find Mendoza (Sidonia) giving the benefits of his wisdom to the King of the French, and as the monarch leaves the room, the Jew whispers to Codlingsby: 'His Majesty is one of *us* . . . so is the Pope of Rome; so is ***'—a whisper concealed the rest.[53] The discreet asterisks clearly hide the name of Queen Victoria, providing a splendid climax to Thackeray's mocking parody of Disraeli's inane hyperbole.[54]

It was Anthony Trollope who created a second Sidonia in *Barchester Towers* (1857). Trollope shared with G. H. Lewes a profound dislike of Disraeli,[55] and an unqualified admiration for Thackeray, so that it is not surprising to find his Sidonia a repellent figure, 'a dirty little old man' (Ch. 9), hounding the 'hero' Ethelbert Stanhope for a debt of £700, which his irate father determines he shall not honour. The episode ends with a cynical apostrophe:

> It is thus, thou great family of Sidonia—it is thus that we Gentiles treat thee, when, in our extremest need, thou and thine have aided us with mountains of gold as big as lions—and occasionally with wine-warrants and orders for dozens of dressing-cases. (Ch. 19.)

Osalez, the stockbroker reflection of Rothschild power and wealth, occasioned no furore from the critics, and failed to survive in the world of literature. His Jewishness was minimized and all but cancelled out by his orthodox Christianity. Further, he did not wish to make the world uncomfortable by trying to re-establish the bases of its continuation: instead, he provided readers with a delicious *frisson* of guilty recognition when he castigated the hypocrisy that pretended to hate money when in fact society was centred upon its acquisition.

[53] Thackeray, *Works*, xv (1879), p. 31.

[54] 'Disraeli's famous roll of European ministers of finance who were Jews, so often quoted by anti-semites, was quite incorrect. None of them was, and in his pride at his descent Disraeli had unknowingly given both here and elsewhere a formidable weapon to the fanatical enemies of his race.' (Blake, *Disraeli*, p. 202.)

[55] Of Disraeli's work he wrote: 'the glory has ever been the glory of pasteboard, and the wealth has been a wealth of tinsel. The wit has been the wit of hairdressers, and the enterprise has been the enterprise of mountebanks' (*Autobiography* (World's Classics edn., 1968), p. 223).

Sidonia, though, was guilty of a new idea—he disturbed preconceptions with his plea for a theocratic government based on his notion of the primitive church. For this reason the critics ignored the fairly obvious links between him and Nathan and Lionel Rothschild, and concentrated instead on reducing him to his 'rightful' place, the old clothes shop and the moneylender's den. This was unfortunate, since, while Sidonia *is* insufferable and monomaniac, his creator made him embody a significant idea, worthy, if not of acceptance, then at least of understanding. What does emerge clearly from a reading of *Coningsby* and *Tancred* is that while the Rothschild links may be clearly seen, Sidonia develops into a figure far removed from any member of that family, and may be described with truth as a projection of Disraeli's fantasies about himself.[56]

It is interesting to see how, in the novels that have been discussed, and indeed in most Victorian novels of finance, it is the client, rather than the financier, who is taken to task for seeking Mammon. Osalez, even in his guise of moneylender, never abandons his clients to total ruin: he is concerned more with their liberation from that spirit of money-getting that had soured much of his own private life. At one stage in *The Moneylender* he makes this point to his client the spendthrift Countess of Winterfield:

> So long as you enjoy luxuries which you do not and cannot pay for, you are shining at the cost of your coach makers, jewellers, milliners, money-lenders . . . Though you may be a Countess of the realm, and I the villified A.O., I rise above you as a capitalist, I rise above you as a moralist . . . (iii, Ch. 1, pp. 8–9.)

The truth of Osalez's assertions in the above passage is morally indisputable, especially as it becomes clear that his reputation for usury has been created by the distorted vision of these social parasites. There is no disgrace in amassing wealth and thus power, provided that this is not done for its own sake, or to establish a deliberate ascendancy over others for selfish purposes.

[56] There is a further Rothschild reflection in Charles Lever's *That Boy of Norcott's* (1869), where the fortunes of an Adriatic merchant depend upon the patronage of 'the great Jew house of "Nathanheimer" of Paris'. The element 'Nathan' in this name is self-explanatory, while the death of Baron James de Rothschild of Paris in 1868 may have also been in Lever's mind. No member of the 'Nathanheimer' family appears in the story, but it is interesting to see the house involved in financing public works, a characteristic of European credit institutions in the sixties, and an activity in which the Rothschilds participated. Lever's word-portrait of Rothschild activity is totally devoid of the mystery and melodrama associated with Osalez and Sidonia.

Similarly Sidonia, intellectual, aesthetic, passionless, achieves a financial pre-eminence that arouses ungrudging admiration in the 'great', such as Lord Monmouth. But he uses his wealth and power for a spiritual purpose, autocratic as it may be, offering ungrudging financial and spiritual aid to the youthful questers Coningsby and Tancred.

These authors, products of a thrusting and confident age, are content to condemn an abstract love of gain; they may fear the power of men like the Rothschilds, but they cannot entirely stifle their admiration. Moral and spiritual conversion in the great is what they seek, rather than any revolutionary revision of the basis of their money-centred society.

6

Financiers (II): John Sadleir and Dickens's Mr Merdle

Dickens's novel *Little Dorrit* appeared in monthly instalments between December 1855 and June 1857. Ostensibly set 'thirty years ago', the date of the action is established in Book I, chapter 18 as 1826,[1] the year of the great commercial crisis. However, two strands of the story—the satire on Government departments as the Circumlocution Office, and the character and career of Mr Merdle—belong more to the forties and fifties, which previous chapters of the present work have shown to be decades of particularly irresponsible commercial profligacy.[2]

Britain embarked upon the war against Russia in the Crimea in March 1854, and from the outset it was clear that an inept bureaucracy was largely responsible for the reverses at Balaclava and Sebastopol, and for the lack of proper equipment and accommodation for the troops. Public outcry led to the fall of the Aberdeen ministry and the accession of Lord Palmerston as premier in January 1855. It was in this month, six months before the first page of *Little Dorrit* was written, that Dickens began his 'Book of Memoranda', which contained many ideas used later in the novel. There is no direct reference in this document to the Crimean War or to the bunglings of administration, and it would be over a year before the Circumlocution Office made its appearance in print. However, it is clear that this deficiency and corruption in the governmental apparatus are the 'political vice' to which Forster referred when he described the various themes of *Little Dorrit* as 'all parts of one satire levelled against prevailing political and social vices'.[3]

[1] 'Here lie the mortal remains of JOHN CHIVERY . . . Who died about the end of the year one thousand eight hundred and twenty-six . . .' (*Little Dorrit* i, Ch. 18).

[2] See in particular Humphry House, *The Dickens World* (1941), pp. 28–9.

[3] Forster, *The Life of Charles Dickens* (1874), iii, p. 136. The substance of this paragraph is drawn from John Butt's 'The Topicality of *Little Dorrit*', *University of Toronto Quarterly*, No. 29 (1959), 1–10. In this article, Dickens's continued pre-occupation with administrative reform both before and during the composition

Miss Mulock's *John Halifax, Gentleman*, which makes the 1826 crisis a principal focus of interest, appeared in 1856, the year in which the greater part of *Little Dorrit* was issued, so that there is a prima facie case for assuming that Dickens, with the same crisis year in mind, meant Mr Merdle to be either a 'bubble' promoter of the period of the Company Mania, or one of the bankers who collapsed at that time. However, there is no need to assume that Dickens had given particular attention to the idea of including a financier in his novel before the opening months of 1856, so that the connection of Merdle with 1826 is at best tenuous.[4] As the chapter progresses, it will be seen how Dickens's financier is the result of several commercial stimuli that belong firmly to the fifties.

There is probably some truth in the idea that Dickens's depiction of Mr Merdle was coloured in some part by recollections of George Hudson, the 'Railway King', whose career is outlined in a later chapter of this book.[5] Hudson had suffered his spectacular fall from power in 1849, and had first appeared in fictional form as Richard Rawlings in Robert Bell's *The Ladder of Gold* (1850); in the following year, thinly disguised as 'Humson', he was a principal character in Emma Robinson's *The Gold-Worshippers*. Nevertheless, Merdle is a very different man from Hudson, gloomy and retiring where Hudson was flamboyant and forceful; a fraud who perishes by his own hand where Hudson was merely a gross mismanager of accounts[6] who went to his grave in 1871 with a crowd of mourners, while the great bell of York Minster was tolled in tribute.[7] When Dickens came to conceive the character of Mr Merdle, spectacular and truly fraudulent financiers

of *Little Dorrit* is examined and illustrated in detail. This is the only article I have seen that appreciates the significance for Dickens of the fall of Strahan & Co.

[4] Book i, Ch. 18, where the date of the story is established, appeared in No. V, Apr. 1856, but was under composition in the latter months of the previous year. All references to dates of composition and issue in this chapter are derived from P. D. Herring's meticulous article 'Dickens' Monthly Number Plans for *Little Dorrit*', *Modern Philology* 64, Aug. 1966, pp. 22–63.

[5] Links between Merdle and Hudson are suggested in, for example, House, *The Dickens World*, p. 29, and Grahame Smith, *Dickens, Money and Society* (1968), p. 165. My dissent from these views will appear as the chapter progresses.

[6] Details of Hudson's financial transactions, and the nature of his 'fall' will be found in Ch. 8 of this book.

[7] See the graphic account of Hudson's funeral, *The Times*, Saturday, 16 Dec. 1871, p. 9.

lay to hand at the time of composition, and there was no need for him to hark back to the events and figures of the forties: 'Mr. Merdle is the man of this time. The name of Merdle is the name of the age.' (ii, Ch. 5.)

In his preface to the one-volume edition of *Little Dorrit* published in 1857, Dickens is careful to dissociate Merdle from the figures of the Railway Epoch, and to place precise limits on the public events that may in part have influenced his depiction of the swindling financier:

> If I might make so bold as to defend that extravagant conception, Mr. Merdle, I would hint that it originated after the Railroad-share epoch, in the times of a certain Irish bank, and of one or two other equally laudable enterprises. Were I to plead anything in mitigation of the preposterous fancy that a bad design will sometimes claim to be a good and an expressly religious design, it would be the curious coincidence that it has been brought to its climax in these pages, in the days of the public examination of late Directors of a Royal British Bank.[8]

The 'certain Irish bank' to which Dickens alludes was the Tipperary Joint Stock Bank, which is examined later in this chapter; it failed in February 1856. The Royal British Bank[9] failed on 3 September 1856, a victim of malversation; it was in that month that Number X of *Little Dorrit*, recounting (in chapter 33) the initial stages of Merdle's decline, appeared in print. It is thus clear that Dickens, quite naturally, was influenced by the 'high art' crimes of the fifties when he came to shape 'that extravagant conception, Mr. Merdle', and felt impelled to suggest this in the preface to the first edition of his novel. Why did he wish to include a fraudulent financier in this work? How important to an understanding of Merdle, and indeed of *Little Dorrit*, is a knowledge of contemporary commercial scandals? What follows will, it is hoped, provide some acceptable answers to these questions.

Dickens had begun his 'Book of Memoranda' in January 1855. In this collection of notes he recorded an idea for allowing Arthur Clennam to suffer a reversal of fortune: 'His falling into difficulty, and himself imprisoned in the Marshalsea. Then she, out of all her

[8] *Little Dorrit* (1857), pp. v, vi.

[9] The closure of this bank had no direct or significant influence on *Little Dorrit*, as the directors were not actually brought to trial until 13 Feb. 1858, a year after the one-volume edition, with its preface, was published. The fullest account of the Royal British Bank is in Morier Evans, *Facts, Failures and Frauds*, pp. 268–390.

wealth and changed station, comes back in her old dress, and devotes herself in the old way.' (Forster, iii, p. 248.) The phrase 'falling into difficulty' suggests that Dickens intended to contrive a financial reversal for Clennam in one of the classical ways: the failure of a speculation, or the fall of a bank, would be the most obvious and realistic device, as there had been innumerable precedents in the commercial life of the century. Although Merdle is one element in Dickens's scheme of social castigation, he is also partly a device for engineering his creator's fancy, early conceived, of depicting Clennam as a debtor in the Marshalsea, and Amy Dorrit as his liberator.

Dickens was engaged on the first chapters of *Little Dorrit* in June 1855. The run upon the defaulting banking house of Strahan, Paul & Bates commenced on Friday, 8 June, and the public announcement of the bank's failure was made on the following Monday. The principals were listed in the *Gazette* as bankrupts on Tuesday, 12 June, and a week later warrants were issued for the arrest of the partners. While these events were taking place, Dickens was writing the first three chapters of Number I,[10] which include Arthur Clennam's dreary homecoming. With Clennam's future perhaps in mind, the closure of Strahan & Co. at this time could certainly have suggested a valid means of effecting his ruin; in any case, this bank was surely one of the 'other equally laudable enterprises' mentioned in his preface.

Work on Number II continued during August 1855, and Dickens began writing Number III in September. On the 18th of this month a true bill was found against Strahan and his partners. Their trial commenced on 26 October, and the severe sentences were passed on the following day. During this month and the succeeding November Dickens was engaged in writing Numbers IV and V. It is in the final chapter of Number V that Dickens allows John Chivery to fix the date of his story as 1826.

In all the chapters so far mentioned, none of which would, of course, be published until December 1855, there is no hint that a figure of Merdle's type is to be introduced as a principal element of the plot. But those characters who are to be radically affected by a reversal of fortune—Mr Dorrit, Clennam, and Doyce—had all been introduced, and the spectacular and fraudulent downfall of Strahan, Paul & Bates, which rendered so many innocent people destitute, could very easily

[10] The fourth chapter, 'Mrs. Flintwinch has a Dream', was added to No. I after No. II had been written.

have suggested to Dickens a topical and relevant means of affecting dramatically the fortunes of these characters. It was after the publication of Numbers I and II that the sensational suicide of John Sadleir, the swindling financier, in February 1856, gave Dickens a more definite and more dramatic model upon which to base Mr Merdle, who was to be both fraud and suicide.

John Sadleir, a Dublin solicitor, came to England from Ireland in 1846, at the height of the Railway Mania. Adopting the lucrative profession of parliamentary agent, he specialized in the guiding of Irish Railway Bills through Parliament. He was accordingly cultivated by the Roman Catholic interest, being returned as MP for Carlow in 1847. It was from this time that his reputation as a great financier developed: he became one of the Messianic figures ushering in the new Age of Gold, securing the almost imbecilic trust of an uncritical public. His patronage was eagerly sought for companies of all types, and he was soon installed as chairman of the Royal Swedish Railway Company, director of the East Kent line, and manager of many other enterprises—'a veritable little Hudson'.[11]

Once established in the public eye as a great financier, Sadleir turned his attention to banking, converting the Tipperary Bank, founded by his grandfather, from private to joint-stock status. For some time his brother James was sole director until a full board could be constituted, so that mismanagement and malversation went unnoticed, James being of much the same mettle as his brother.

Sadleir's public career continued to prosper. It was suggested in government circles that he could one day be Chancellor of the Exchequer, and when the ultra-Protestant Lord Aberdeen took office in 1853, Sadleir was easily persuaded to forsake the Catholic interest. Consequently rejected by the Carlow electorate, he was, by rather devious means, returned as MP for Sligo. He was soon appointed a junior Lord of the Treasury, and it was at this high point in his career that things began to go wrong. Rumours began to circulate that there were irregularities in his business ventures, and he was curtly asked to resign. His resignation caused questions to be asked in the City, and finance houses with whom he dealt began to insist on punctual meeting of drafts due in cash.

[11] Morier Evans, *Facts, Failures and Frauds*, p. 228. The brief account of Sadleir given here is derived in the main from Ch. vi of this work.

On Thursday, 14 February 1856 Glyn & Co., the London agent-bank for the Tipperary Joint-Stock Bank, refused to honour the bank's drafts, as no remittances had been paid in. Sadleir made frantic efforts to remain solvent, inserting explanatory statements in the press, and approaching various bankers with schemes which they refused to accept. One such proposal, made to the celebrated finance house of Wilkinson, Gurney, Stevens & Co., was so peculiar that suspicions were aroused, and the firm made it clear to Sadleir that it intended to re-examine his securities, which were in the form of deeds seemingly under the seal of the Encumbered Estates Commission.

This encounter with Wilkinson, Gurney was the breaking-point. Sadleir knew that these deeds, and the signatures appended to them, were forgeries, and that the seal, though genuine, had been fraudulently transferred from another document. No parliamentary immunity covered such grave felonies, and Sadleir was well aware that he now faced many years of penal servitude or transportation: it was only four months previously that Strahan, Paul, and Bates had been so sentenced. Returning to his house on the evening of Saturday, 16 February, he began the preparations for his terrible end.

Between 11.30 p.m. and midnight on Saturday, 16 February, Sadleir left his house in Gloucester Square without informing his domestic staff.[12] At about twenty minutes to nine on the following morning, his body was found by a labourer on Hampstead Heath. The discovery of the body was reported in *The Times* on Monday, 18 February 1856:

> The body of Mr. J. Sadleir, M.P., was found on Sunday morning, February 17, 1856, on Hampstead Heath, at a considerable distance from the public road. A large bottle, labelled 'Essential oil of bitter almonds', and a silver cream-jug,[13] both of which contained a small quantity of the poison, lay by his side.

The wide-ranging defalcations of Sadleir were soon revealed. The Tipperary Bank had suspended payment, and all kinds of title-deeds, securities, and acceptances were found to be forged. Sadleir and his brother had cheated the bank of £200,000, and the house failed for

[12] Unless otherwise acknowledged, the account given here is derived from the reports of the three coroner's inquests on Sadleir, held on 19 and 25 Feb. and 11 Mar. 1856. Dickens was engaged on the sixth number of *Little Dorrit* during Feb. and Mar.; he began work on the seventh in Apr. The coroner's reports are reproduced in Morier Evans, *Facts, Failures and Frauds*, p. 240 f.

[13] This is, of course, the cream-jug mentioned so cruelly and callously by Dickens in his letter to Forster (see note 15).

£400,000. The greatest single forgery committed by Sadleir was his issue of false shares to the nominal value of £150,000, the proceeds of which fraud he appropriated to his own use. It need hardly be mentioned that numerous innocent investors and depositors were utterly ruined by this man's criminal acts.

Number III of *Little Dorrit*, containing chapters 9, 10, and 11, was published in February 1856, and on the 19th,[14] two days after Sadleir's suicide, Dickens sat down to write Number VI, which was due for publication in May. Number VI was to cover chapters 19 to 22, and it is in the last two chapters of this number that the Merdles make their appearance. In a letter to Forster Dickens declared: 'I had the general idea of the Society business before the Sadleir affair, but I shaped Mr. Merdle himself out of that precious rascality . . . Mr. Merdle's complaint, . . . fraud and forgery, came into my mind as the last drop in the silver cream-jug on Hampstead heath.'[15]

Thus Sadleir's death, occurring only two days before he began writing this Number, provided Dickens with an excellent, immediately contemporary model for the kind of commercial figure foreshadowed in the 'Book of Memoranda' of 1855 as a means of involving Clennam in 'difficulty'. A suicide is more effective dramatically than a transportation, and Sadleir is thus a better prototype than Dean Paul, though of course the damage done by both men, and the moral to be drawn from their respective fates, are identical.

It is Mrs Merdle who first informs the reader of the nature of her husband's business. 'Mr. Merdle is a most extensive merchant, his transactions are on the vastest scale, his wealth and influence are very great . . .' (i, Ch. 20). Her words set the tone of hyperbole that will characterize any reference to Merdle's financial activities before his fall. In chapter 21 we learn that he is 'immensely rich, a man of prodigious enterprise', and in the course of the conversation at his dinner party in Harley Street in the same chapter, he is said to have made 'another enormous hit', possibly involving four hundred thousand pounds.

We are told nothing more than this of Merdle's business in the chapter that introduces him, but the reader who saw this chapter in the sixth number in May would have remembered how John Sadleir

[14] This dating is derived from Butt, p. 8.
[15] The letter is reproduced in Forster, iii, p. 136.

had been spoken of in similar terms. Dickens was paving the way for a crash of a type that was hardly unfamiliar to his readers.

As Merdle's influence over the various societies within the novel increases, so the vague, unreasoning hyperboles are sustained.

> Mr. Merdle came home, from his daily occupation of causing the British name to be more and more respected in all parts of the civilised globe, capable of the appreciation of world-wide commercial enterprise and gigantic combinations of skill and capital. (i, Ch. 33.)

In the sentence following this quotation, however, Dickens reveals the irony underlying this unthinking adulation when he remarks that 'nobody knew with the least precision what Mr. Merdle's business was . . .'.

Artistically it is unnecessary for Dickens to show Merdle at work: contemporary readers could draw analogies from similar figures in real life, particularly Sadleir, and also from fiction of the period.[16] A more important reason for this deliberate vagueness is the need to emphasize the mesmerizing powers of these magnates over the investing public, and the spread of a madness of credulity to all ranks of society. Nobody knew what Merdle did; everybody knew that he was the greatest man of the age.

When Dickens reached the twelfth chapter of Book II he linked Merdle specifically with banking, because it would be through a classic and credible bank failure that the principal characters of *Little Dorrit* were to be ruined. In the general note for his number plan for Number XII Dickens had written: 'Pave the way—with the first stone—to Mr. Merdle's ruining everybody? *Yes*.'[17]

Number XIII moved away from Merdle, but Number XIV continued to 'pave the way'. In his number plan for chapter 12 Dickens wrote the phrase 'The Wonderful Bank', and accordingly in this chapter Merdle is linked specifically with banking: 'the evening paper was full of Mr. Merdle. His wonderful enterprise, his wonderful Bank . . . . The wonderful Bank, of which he was the chief projector,

[16] In Apr. 1857, when Mr Merdle's suicide appeared in print, the London and Eastern Banking Corporation, which had been established as a joint-stock enterprise in Jan. 1855, failed, the victim of a particularly impudent fraud perpetrated by some of its directors and managers.

[17] Herring, p. 44. The word 'Yes' is underlined five times.

establisher and manager,[18] was the latest of the many Merdle wonders.' (ii, Ch. 12.)

Having created an atmosphere of unreasoning trust in the probity of a financier unconfirmed by factual knowledge, and having involved that financier in the currently shaken world of banking, Dickens proceeds, in Book II, chapter 13, to present an epitome of the kind of speculative madness that was a characteristic precursor of the familiar crash.[19] Particularly affected are the denizens of Bleeding Heart Yard, and it is Mr Pancks who most vividly illustrates the debilitating effect of the social mesmerism engendered by universal trust in Merdle. Clennam remarks to Pancks how very strange it is that 'runs on an infatuation prevail', and he agrees, declaring that such runs arise from general ignorance of money matters. Pancks does claim to understand money, and the reader remembers that he had been largely instrumental in securing Mr Dorrit's fortune for him.

It is thus very effective to learn, a few paragraphs after this exchange with Clennam, that Pancks has actually parted with £1,000 to Merdle, and that he justifies his trust by mouthing the tell-tale hyperboles that show a judgement warped by the popular self-deception: 'I tell you, Mr. Clennam, I've gone into it . . . He is a man of immense resources—enormous capital—government influence. They're the best schemes afloat. They're safe. They're certain.' (ii, Ch. 13.)

Arthur Clennam instinctively shies away from the idea of speculation; but Pancks, a man whom he both trusts and highly regards, is persuasive, and ironically succeeds in swaying Clennam to his way of thinking by a second recommendation that is virtually a verbatim repetition of his original remarks:

> 'But what of Go in and lose?' said Arthur. 'Can't be done, sir,' returned Pancks. 'I have looked into it. Name up, everywhere—immense resources—enormous capital—great position—high connexion—government influence. Can't be done!' (ii, Ch. 13.)

But, of course, it can be done. Mr Merdle kills himself, and immediately the 'wonderful bank' closes. It is realized that he

> had been, after all, a low, ignorant fellow; he had been a down-looking man, and no-one had ever been able to catch his eye . . . he had never had any

[18] Dickens would have known that the Tipperary Bank had been, in effect, the personal property of Sadleir and his brother, a mockery of its joint-stock character.

[19] Dickens had provided a more specific, localized treatment of the same theme earlier in *Nicholas Nickleby* (1838), where his account of the ruin of Nicholas's father shows a detailed awareness of the Crisis of 1826.

money of his own; his ventures had been utterly reckless, and his expenditure had been most enormous. (ii, Ch. 25.)

It is in this effective manner that Dickens shows a mesmerized society coming abruptly out of its trance, looking behind the distorting superlatives to the commonplace reality.

A writer of Dickens's stature did not need any specific commercial event of the fifties to suggest themes and subjects for his novel.[20] The criminal frauds specifically mentioned in this chapter and elsewhere would have contributed in some measure to the author's creative imagination, but perhaps of equal importance was the general progress of commercial activity during the decade, both at home and abroad.

In 1848 valuable gold deposits were discovered in California, and in May 1851 massive deposits were found in New South Wales. Following both discoveries there was the expected wide-spread agitation in trade, and a perilous expansion of credit. Very soon there developed a mania for gold-mining companies, both in England and America, many of which failed, being inept, ill-conceived, or downright fraudulent. Monthly reports of vast quantities of gold arriving at Port Philip and Sydney for shipment caused further expansion of credit at home, and an enormous increase in all kinds of trading. The total value of exports in 1848 was £60 million; in 1857 the figure had increased to £122 million.

A crash was, of course, inevitable,[21] and this time it was triggered off by events in America. On 24 August 1857 the Ohio Life Insurance Co. of Cincinnati and New York failed, with liabilities of seven million dollars. This failure signalled the almost total collapse of American commerce, which had for nearly a decade accommodated itself to under-capitalized banks, an excess of currency, and a gross extension of discounting. In all, 5,123 failures occurred in the USA and Canada, with liabilities of nearly 300 million dollars.[22]

A tremendous convulsion was felt in England, where massive failures of well-respected banks revealed much gross mismanagement and

[20] 'It is not that Dickens required the notorieties of the mid eighteen-fifties to provide him with episodes and characters for his novel; it is rather that he had already taken imaginative stock of the situation when some fresh event occurred to confirm his diagnosis and to supply him with an illustrative example'. (Butt, p. 8.)

[21] 'a repetition will occur at future periods, in some shape or other, despite every precaution that may be attempted . . .' (Morier Evans, *Commercial Crisis 1857*, p. 27).

[22] See Morier Evans, *Commercial Crisis 1857*, p. 34; and A. Andréadès, *History of the Bank of England* (1909), pp. 346–7.

fraud, and only Government intervention saved the day. Morier Evans rightly remarked that, of all commercial crises, that of 1857 was 'the most severe that England, or any other nation, has ever encountered.'[23]

These events formed part of Dickens's contemporary awareness during the two years of the composition of *Little Dorrit*. Although the crisis of 1857 occurred long after the novel was finished, Dickens, in common with any intelligent observer, could not have failed to see where the monumental increases in trade and credit would lead.

There are one or two expressions used by Dickens in describing Merdle's commercial reputation that suggest a further contemporary financial development influencing the author's depiction of his entrepreneur. Mrs Merdle tells us that 'his transactions are on the vastest scale' (i, Ch. 20), and Dickens himself links him with 'world-wide commercial enterprise and *gigantic combinations of skill and capital*'. (i, Ch. 33. My italics.) To the student of nineteenth-century commercial history these expressions, and particularly the words italicized, suggest that Dickens was alluding to the vast French joint-stock company known as the Crédit Mobilier.

During the greater part of 1856 Dickens and his family resided in Paris, at a house in the Champs Élysées. It was four years since the establishment of the Second Empire, and an optimistic, excitable society indulged itself in luxurious sybaritic display. Dickens was fêted by the various intellectual and theatrical coteries, and regaled Forster with detailed accounts of his entertainment.[24]

He had, though, time to notice and to comment upon the frantic activity centred on the Paris stock exchange, the Bourse. 'Instant prosperity' was the order of the day, and fortunes were made and lost with a rapidity that put similar English transactions in the shade. At a banquet given in his honour by Émile de Girardin, Dickens was shown a man who typified the inflated gambling spirit of the Empire.

> A little man dined who was blacking shoes 8 years ago, and is now enormously rich — the richest man in Paris — having ascended with rapidity up the usual ladder of the Bourse. By merely observing that he might come down again, I clouded so many faces as to render it very clear to me that *everybody present* was at the same game for some stake or other![25]

[23] Morier Evans, *Commercial Crisis 1857*, p. [v].

[24] Forster, iii, Ch. 5, *passim*.

[25] In a further letter to Forster (iii, p. 119), Dickens describes the scene at the Bourse, with crowds of speculators 'all howling and haggard with speculation'. It is quite possible to see Rigaud, the 'smooth, polished scoundrel' of *Little Dorrit*, as a

With this awareness of commercial activity it is more than likely that Dickens would have heard talk of the greatest single speculative venture of the decade, the Crédit Mobilier.

This undertaking was a joint-stock company, organized with limited liability in November 1852. Its capital, equivalent to £2,400,000, was apportioned in shares of £20 each. By means of this capital, and through money raised by credit, the company launched into a seemingly limitless series of purchases: Clapham remarks that its object was 'the development and encouragement of everything'.[26] Shares were purchased in factories, shipping lines, railways, omnibus companies, with a view to consolidation and rationalization; paper mills, canal companies, and building land were soon added. Further, the company projected and built hotels, theatres, and other amenities, creating public demands which it could then supply.[27] It was a mammoth undertaking, representing, like the Merdle enterprises, 'gigantic combinations of skill and capital'.[28]

While the fever of speculation on the Bourse maintained its impetus, the Crédit Mobilier did well, particularly as many prominent financiers associated themselves with its dealings, and made large profits through high rates of interest and commission on stock transfers. This great umbrella organization carried out many functions normally found in a number of individual enterprises. It dealt in millions, and undertook obligations, many of a constructional character, not only in France, but in Switzerland, Spain, Austria, and Russia. Its very omnipotence

type of the unprincipled opportunists thrown up by the Second Empire, or, indeed, as a canard upon the Emperor himself. In his review of No. I in the *Athenaeum* (1 Dec. 1855, pp. 1393–5), W. Hepworth Dixon wrote of 'the magnificent French adventurer, Monsieur Rigaud. This latter personage, we suspect, is not unlikely to be a prominent actor in the tale; and he is altogether so magnificent a creature (and is so like a well-remembered criminal of real life) that we cannot doubt the reader's desire to make his acquaintance.' (p. 1394). I have not identified the criminal to whom Dixon refers.

26 Clapham, *Economic History*, ii, p. 364.

27 Morier Evans, *Commercial Crisis 1857*, p. 40.

28 The company saw itself as a universal banking and brokerage institution. Further, it had contemplated creating a new form of circulating paper, similar in function to an English Exchequer Bill, which was to be used instead of the Imperial currency to discount bills of exchange and other negotiable paper. See in particular W. Newmarch, 'On the Recent History of the Crédit Mobilier', *Journal of the Statistical Society*, Series A, xxi, Dec. 1868, p. 444 f. See also L. Levi, *History of British Commerce*, 2nd edn., Ch. 5, p. 461, where the company's objects are further described. Levi states that the value of the company's circulating paper was £24 million.

was one of the reasons why Paris held out so many attractions as a capital market in the 1850s.

The company was not actively dishonest, but in its leviathan constitution lay the seeds of its demise. Profits diminished during 1856 and 1857, when the French speculative mania began to abate, and it became clear that the grandiose principles of the Crédit Mobilier were confusing, ill-defined, and commercially ephemeral: 'we find it gradually crumbling away; principle failing after principle, one mode of practice abandoned after another, this investment and that rapidly giving way, till at the present time there is good reason to believe that it is in the last stages of its existence.'[29]

Morier Evans states that wealthy business men who had become entangled with the company over-speculated on the Bourse and fled to America, owing hundreds of thousands of pounds. Frauds were practised in connection with some of the company's railway and dock undertakings, and it is possible that 'exalted Government authorities' were involved in these frauds.[30] The gigantic fabric began to wear thin after 1856. In 1857 no dividend was paid, and ten years later the ill-starred company failed. It is, I believe, valid to suggest that the operations of the Crédit Mobilier, with its fund of superlatives, may have formed part of Dickens's commercial awareness when he created Mr Merdle.

Mr Merdle was to be something more than a mere reflection of Sadleir. It is clear, both from the text of the novel and from the number plan for Part VI, that both Mr and Mrs Merdle had been conceived as an indictment of that hollow Victorian concept, 'Society', which so exercised the talents of Mrs Gore, and which Dickens himself had earlier satirized in the person of Mrs Wititterly in *Nicholas Nickleby* (1838–9).[31] Chapter 20 is entitled 'Moving in Society', and in his number plan for this chapter Dickens had written: '*Mrs. Merdle*—Bosom—*Society, Society, Society.*'[32] The theme is continued in chapter 21, 'Mr. Merdle's Complaint', the number

[29] Newmarch, p. 451.

[30] Morier Evans, *Commercial Crisis 1857*, p. 41.

[31] Dickens's clever parody of a 'silver fork' novel, 'The Lady Flabella', is in Ch. 28 of *Nicholas Nickleby*. Kathleen Tillotson, in her *Novels of the Eighteen Forties* (1961), pp. 73–4, shows that it is an exuberant mocking of the style of Bulwer, Hook, and Disraeli, who all wrote novels of 'high life'. I disagree with Mrs Tillotson's inclusion of Mrs Gore as a subject of Dickens's satire: Mrs Gore was herself a satirist not only of the mode, but of the society that loved to read the novels that belonged to it.

[32] Dickens underlines 'Mrs Merdle' twice, and 'Society, Society, Society' three times.

plan for which contains as its second note: 'Still Society—always Society—Everything for Society.'[33]

It is the gods of capital, personified by Merdle, who seem to be worshipped with such unthinking adulation by all strands of this 'Society'. Those who pander most to Merdle are those who, from their several stations in the social hierarchy, secretly despise him: 'True, the Hampton Court Bohemians, without exception, turned up their noses at Merdle as an upstart; but they turned them down again, by falling flat on their faces to worship his wealth.' (i, Ch. 33.)

The Victorian reader would find nothing in Merdle's career and end that could not easily have been guessed at in the context of contemporary commercial life. Dickens is more concerned to show the moral sickness of society than he is to condemn Merdle and his kind. This is why Merdle is so skilfully underdrawn. He makes little noise about himself, but those around him puff him to the skies.

> Nobody knew that the Merdle of such high renown had ever done any good to anyone . . . All people knew . . . that he had made himself immensely rich; and, for that reason alone, prostrated themselves before him, more degradedly, and less excusably than the darkest savage creeps out of his hole in the ground to propitiate, in some log or reptile, the Deity of his benighted soul. (ii, Ch. 12.)

Merdle appears to have done or said nothing to persuade people to worship him: 'he had not very much to say for himself' (i, Ch. 21). The impetus of greed and moral confusion impelled society not only to flock to Merdle, but to praise and flatter him with inane compliments: it is they, mesmerized by the 'enchanted name',[34] and not Merdle, who form the principal focus of Dickens's trenchant condemnations.

Dickens saw this blindness to reality, engendered by cupidity, as 'a moral infection' that would spread 'with the malignity and rapidity of the Plague' (ii, Ch. 13), and it is important to realize that the Plague strikes all alike. Merdle's defalcations are side-effects of the one disease

[33] Mr Herring writes of this number, and of these two chapters in particular: 'The Society in which the Merdles move constitutes the last and most comprehensive in the series of prisons presented in the novel . . . Mr. Dorrit was to escape the Marshalsea and the clutches of his creditors in the Circumlocution Office only to find himself in a social world equally restricting. And equally perfidious, for the most respected figure in Society would be guilty of "Fraud and Forgery bye and bye".' (Herring, p. 35.) 'Fraud and Forgery bye and bye' is the last entry in the number plan for Ch. 21.

[34] Dickens uses this expression in his number plan for No. XIV, Ch. 13 (Herring, p. 48).

that is afflicting society: he is both created and destroyed by it. Mr Merdle's mysterious 'complaint', which turns out to be fraud and forgery, is but one symptom of a disease common to all.[35]

Just as Society is mesmerized by Merdle, so too is Merdle mesmerized by Society. Merdle and his wife have been 'licensed' by Society to move in the *haut monde*: 'Society was aware of Mr. and Mrs. Merdle. Society had said, "Let us license them; let us know them".' (i, Ch. 21.) Merdle, far from enjoying or exulting in this worship, is rendered personally miserable by it, but Society's flattering seduction makes him strive to defer to it in all things. He may deceive others financially, but is himself utterly deceived into thinking that the hollow sham which he propitiates is something of value and esteem: '. . . his desire was to the utmost to satisfy Society (whatever that was) . . . [He was] tenacious of the utmost deference being shown by everyone, in all things, to Society . . .' (i, Ch. 21.)

Merdle is a reserved, modestly-spoken man, not fond of company, and plainly ill at ease when entertaining. It is his 'duty' to Society that makes him endure public life, and at many points in the novel Merdle's endurance of inward discomfort for the sake of Society is carefully, and sometimes amusingly, described.[36] Only twice in the novel does this spiritless, vapid man show signs of emotion, and the two episodes in which he does so hold special significance for our understanding both of his character and of his role in the novel.

In Book I, chapter 33, Mrs Merdle tells her husband that his retiring nature renders him 'unfit to go into Society'. Mr Merdle's response to these words is physically and verbally violent. He seizes his hair in his hands and cries:

> Why in the name of all the infernal powers, Mrs. Merdle, who does more for Society than I do? Do you see these premises, Mrs. Merdle? Do you see this furniture, Mrs. Merdle? . . . Do you know the cost of all this, and who it's all provided for? And yet will you tell me that I oughtn't to go into Society? I, who shower money upon it in this way?

[35] See in particular Grahame Smith's perceptive remarks on this interpretation in *Dickens, Money and Society* (1968), pp. 116–7.

[36] See in particular the descriptions of Merdle's general deportment in Book I, Chs. 12 and 21. Merdle's shyness of Society is interestingly paralleled by that of Dudleigh, the 'ruined merchant' in Samuel Warren's *Diary of a Late Physician* (1838). Dudleigh flies to the house of 'some sedate city friend' on the appearance of his wife's fashionable guests (Warren, *Works*, 1854, i, p. 232).

This initial reaction to his wife's words shows how Merdle equates Society with money and luxurious surroundings, and exhibits his rage at the suggestion that he cannot accommodate himself to Society's rituals. But his words do not really account for the untypical anger that informs both his actions and his delivery. The true reasons for his outburst are revealed in his subsequent assertions:

> 'But to tell me that I am not fit for it after all I have done for it—after all I have done for it,' repeated Mr. Merdle, with a wild emphasis that made his wife lift up her eyelids, 'after all—all!—to tell me I have no right to mix with it after all, is a pretty reward.'

This *cri de cœur*, with its accompanying 'wild emphasis', shows that Mrs Merdle's criticism has inadvertently touched a raw nerve, and released, momentarily, a manifestation of Mr Merdle's inner torment. The tell-tale cry of 'After all—all!' reveals that the financier is alluding to the sacrifice of his own retiring nature to the public show required by Society, and, more important, to the guilty knowledge that in his 'all' is included fraud and forgery. The outburst is that of a man who believes in 'Society', and who yet can only remain within it by cheating and defrauding it. 'You don't know half of what I do to accommodate society. You don't know anything of the sacrifices I make for it.' One of the sacrifices is that of whatever honour and personal integrity he may once have possessed.

This is a devastating indictment of Mammon and its influence to corrupt; Merdle the fraud exists only because Society sanctions him. Conversely, Merdle's hypothetical millions sustain that Society in all its empty amorality, and in striving, through fraud and malversation, to create his mythical wealth, Merdle has destroyed not only his personality, but his soul: 'Mr. Merdle, left alone to meditate on a better conformation of himself to Society, looked out of nine windows in succession, and appeared to see nine wastes of space.'

In chapter 16 of Book II, Mr Merdle calls upon Mr Dorrit, with the intention of encouraging him to invest his fortune in his enterprises. We are given no details of Mr Merdle's financial difficulties, but his demeanor in this episode suggests that whatever processes Dickens imagined as effecting his downfall were already in train. Mr Dorrit, fully affected by the general sickness, is highly honoured by Mr Merdle's visit, and listens attentively to his blandishments. Both men, we remember, are devoted to Society. Dickens is careful to describe Merdle's appearance as he speaks to the erstwhile Father of the Marshalsea:

There were black traces on his lips where they met, as if a little train of gunpowder had been fired there . . . Mr. Merdle turned his tongue in his closed mouth—it seemed rather a stiff and unmanageable tongue—moistened his lips, passed his hand over his forehead again, and looked all round the room . . .[37]

This is an admirable portrayal of a man who is, with herculean self-control, concealing acute *fear*, and the closely-observed description adds an element of dramatic tension to the episode. The reader senses that Merdle is making a final desperate attempt to secure money in order to stave off the unthinkable. Merdle, of course, must exhibit no outward signs of agitation; the signs and motions are all within, a part of this man's destruction of himself.

After these two self-revealing episodes, Merdle relapses into his usual vapidity, and nothing else, not even the preparation for his own death, moves him in the same way. Society was all that mattered, and the securing of money by whatever means to keep the approbation of Society, was paramount. The suicide of John Sadleir on the night of 16 February 1856 gave Dickens the means of depicting Merdle's death. He too would take his own life, though not by drinking bitter almonds, as Dickens had already used this device earlier in *Martin Chuzzlewit* (1843).[38] In Book II, chapter 24, Mr Merdle borrowed a penknife from Fanny Dorrit, and it was with this implement that he killed himself. When the police constable who was fetched to see Sadleir's body on Hampstead Heath searched through the dead man's pockets, he found, among other things, 'a small pocket paper-knife', and 'a case containing two razors',[39] and it seems clear that the wretched man had intended to use these implements had the poison not taken effect. The inquests were fully reported in the press, and one ventures

[37] Earlier, he was depicted as 'passing his great hand over his exhausted forehead'. Later in the chapter we are told of 'New passages of Mr. Merdle's hand over his forehead.' I feel that Dickens's insistence on this detail reflects the similar action of John Sadleir in the final days of his life. Mr Anthony Norris, a witness at the first inquest on 19 February 1856, said: 'He did not complain of his head, but I have noticed him put his hand to his head as if he was oppressed' (Morier Evans, *Facts, Failures and Frauds*, p. 246).

[38] Jonas Chuzzlewit used the same poison to kill himself. 'Happening to pass a fruiterers on the way . . . one of them remarked how faint the peaches smelt . . . "Stop the coach! He has poisoned himself! The smell comes from this bottle in his hand!"' (*Martin Chuzzlewit*, Ch. 51). Essential oil of almonds is obtained both from the seed of the bitter almond and the kernels of peaches: the virulent poison hydrocyanic (prussic) acid is found in the essential oil.

[39] Inquest report (Morier Evans, p. 242).

to suggest that Dickens had mentally filed this detail of Sadleir's effects, thus making Merdle employ a means of self-destruction that poor Sadleir had contemplated as a second possibility.[40]

Immediately after Merdle's death, rumour circulated that he had left a confession: 'He had left a letter at the Baths addressed to his physician, and his physician had got the letter, and the letter would be produced at the Inquest on the morrow, and it would fall like a thunderbolt upon the multitude that he had deluded.' (ii, Ch. 25.)

A witness at the first inquest on Sadleir, Mr Anthony Norris, a solicitor, had received a letter from Sadleir posted on the evening of his suicide. Norris was unwilling to put this letter in as evidence, but was ordered by the coroner to do so. It was produced and read at the second inquest, held on 25 February 1856:

> Saturday Night
>
> I can not live—I have ruined too many—I could not live and see their agony—I have committed diabolical crimes unknown to any human being. They will now appear, bringing my family and others to distress—causing to all shame and grief that they should ever have known me . . .

This, and similar letters of a distressing nature were produced at the inquest, and published in the press, and perhaps Dickens conceived the rumour of a Merdle letter from reading them. However, he deliberately leaves the matter of Merdle's letter undeveloped. It would have been artistically wrong to arouse any kind of sentiment or sympathy for Merdle, as he was to function to the last as a symbol of financial abuse, and as a myth fostered upon society by itself. The wretched Sadleir exhibited great agitation in his last hours, but Merdle approached his self-destruction calmly and almost apologetically. His commercial designs—whatever they were—having failed, he had placed himself beyond the pale of that Society which had claimed his unreasoning allegiance; and without that Society life as he understood it was no longer a viable proposition. In one sense, when Mr Merdle borrowed the tortoise-shell penknife from Fanny Dorrit, he was already dead.

[40] Mr Merdle drank laudanum prior to cutting his veins, presumably to induce rapid stupor.

7

Financiers (III): Trollope's Augustus Melmotte and the Corruption of the Old Order

Anthony Trollope's bitter novel *The Way We Live Now* appeared in twenty monthly parts from February 1874 to September 1875; a two-volume edition was issued in the same year. Trollope's first biographer, T. H. S. Escott, described the novel as a satire on 'the idolatry of the golden calf'. Escott linked Trollope's purposes with those of the great editor of *The Times*, William Delane, who had printed a signed denunciation of a 'Californian colonel, long since kicked out of all decent company', who had swindled 'an English nobleman of ancient title and descent' out of £10,000. He suggested that this 'swindling Midas' had given Trollope 'more than a hint' for Melmotte, though he had earlier admitted that Delane's article had appeared too late to have had any fundamental influence on the delineation of Trollope's financier. The truth is that both Delane and Trollope were reacting independently to prevailing abuses. 'With equal disgust and indignation both observed the acceptance of mere wealth as a passport to the company of men who were social leaders by right of birth . . .'[1]

Thomas Escott was born in 1844, and his comments here are a telling reflection of the high Victorian disdain for those many figures of the commercial world who were slowly but inevitably consolidating their victory over the landed interest. No one wishes to justify Melmotte or his frauds, but it cannot be doubted that some of the invective hurled against him arises from his being an interloper, one who has no 'right of birth' to occupy a leading position in society. The most abusive language that Trollope uses to describe him is 'the horrid, big, rich

[1] This and the preceding quotations are from T. H. S. Escott, *Anthony Trollope* (1913), pp. 296 f. Michael Sadleir, in his *Trollope: a Commentary* (1927), suggests that the novel was impelled partly by Trollope's fear of becoming *passé*: 'The critics were beginning to contrast his leisured comedies of county manners with the glorious actuality of business and efficiency' (p. 399).

scoundrel'; Escott calls him 'a grotesque and nauseating monstrosity . . . The bloated and ferocious plutocrat . . .'.[2]

Novelists were quick to condemn the New Men of commerce as unscrupulous, coarse, bloated, probably dishonest, and given to what Bulwer called 'scarlet vulgarity', and the casual reader of *The Way We Live Now* will find many familiar echoes of this literary tradition. Thus anyone who turned from a reading of *Little Dorrit* to a perusal of Trollope's novel would be forgiven for assuming that Melmotte was an echo of Dickens's Merdle, as both fictional characters operate crookedly in a corrupt society, fail financially, and seek escape in suicide. We know that Trollope had read *Little Dorrit*, and was thus familiar with Merdle,[3] but there is no need to regard Melmotte as an imitation of Dickens's financier. Rather we may see how both derive, at least partly, from a common commercial stimulus.

It was suggested in the previous chapter that Merdle and his 'gigantic combinations of skill and capital' owe something to the vast expansion of credit enterprises in the French Empire, and notably to the gargantuan Crédit Mobilier. Limited liability was established in England in 1855, and this provided an immense stimulus for new joint-stock enterprises, as now their losses could never exceed the amount which the investors at the time intended to risk. Of particular importance were the new finance companies that made their appearance in the late fifties and early sixties, many of which aimed at vying with the grandiose operations of the French companies. Among these English houses one may mention the International Financial, the London Financial, the Imperial Mercantile Credit, Crédit Foncier et Mobilier, and the Joint-Stock Discount.

In the same way as the French Crédit Mobilier, these credit houses began to finance public works, but the way in which they did so was dangerous and irregular. Before their advent no public work, such as a railway or a dock, could be undertaken until shares were actually sold and sufficient capital obtained for the purpose. Now, promoters did not need to wait for capital. Projects involving millions could be

[2] Escott, p. 297.

[3] Escott had heard Trollope tell Sir Henry James that he had not read *Little Dorrit* until 1878 (Escott, p. 298). Sadleir accepted this explanation, and dismissed any similarity between Merdle and Melmotte as 'therefore pure coincidence' (Sadleir, p. 400). However, Mr Bradford A. Booth has proved conclusively that Trollope read *Little Dorrit* as it appeared in monthly parts, and that he had submitted an article on the third number to the *Athenaeum*. See Bradford A. Booth, 'Trollope and *Little Dorrit*', in *The Trollopian* 2 (1948), pp. 237–40.

commenced on promises, and the new credit companies were happy to endorse the projectors' bills, stock, and preference shares. Paper of this type was mere speculation on the future. There could be no guarantee that a project would be finished, or that it would make a viable profit: a crisis, involving a collapse of credit, would render all such paper worthless.

By 1866 there were three hundred such companies, with an aggregate paper capital of £504 million, many of which were to disappear wretchedly through abandonment, winding-up, or bankruptcy. The increase in the number of public works financed caused a large part of the floating capital available in the country's banks to become fixed, so that less money was available for normal trade demands. Inevitably, restraint had to be applied through higher interest, and in September 1864 the Bank fixed its rate at 9 per cent, the highest known that century. Excessive demand for capital never ceased, and on 'Black Friday', 11 May 1866, the inevitable crash was signalled by the closure of the distinguished credit house of Overend, Gurney & Co.[4]

This firm of bill brokers and discounters, known as 'the bankers' bankers', had enjoyed the highest reputation for sixty years; it afforded credit facilities second only to those of the Bank of England. Internal rearrangement, and transfer to limited liability, could not rectify increasing difficulties, and the failure of this house triggered off a most severe convulsion. As usual, disaster was averted by a Government letter, in which it was ordered that the interest rate must be raised to 10 per cent.

This, briefly, was the prevailing financial climate during the decade before Trollope planned his novel, and created Augustus Melmotte. After 1866 the ten-year pattern of panic and crisis was broken: the events of that year were followed by twenty-four years of continuing prosperity.[5] Thus Trollope in 1875 is not influenced by the familiar theme of earlier decades: the speculator on the make, his gullible dupes,

[4] This brief account of English credit companies and the crisis of 1866 is derived principally from Levi, op. cit. (2nd edn.), Ch. 9. See also Andréadès, Ch. 3, and the graphic account of panic in the City in *The Times*, Saturday, 12 May 1865. Also on 'Black Friday' occurred the suspension of the firm of Peto and Betts, the great public contractors, who had built the Reform Club and Nelson's Column, and railways in many countries. The firm had developed the port and resort of Lowestoft, and had built the Royal Hotel there, which seems to be described in Trollope's novel (i, Ch. 42, 46).

[5] The last great crisis of the century, known as the 'Baring Crisis', occurred in 1890. See, e.g. Andréadès, Ch. 4, pp. 362 f.

and the general ruin following a panic. He is, rather, keenly aware that credit and credit financing, coupled with the feverish company promotion following the establishment of limited liability, have become acceptable to the higher ranks of society, and that the class from which Britain had for centuries derived its moral and social values was affected and corrupted by the new order of things.

That Trollope had thought carefully about the rapidly advancing importance of credit in commerce is clear from Melmotte's assertions on this subject.[6] To Paul Montague, who objects to the 'rubber stamp' nature of Board meetings, Melmotte explains:

> Look here, Mr. Montague. If you and I quarrel in the Board-room, there is no knowing the amount of evil we may do to each individual shareholder . . . Gentlemen who don't know the nature of credit, how strong it is,—as the air,—to buoy you up; how slight it is,—as a mere vapour,—when roughly touched, can do an amount of mischief of which they themselves don't in the least understand the extent.[7]

Later in the novel he says much the same thing to Lord Nidderdale:

> you don't understand how delicate a thing is credit. They persuaded a lot of men to stay away from that infernal dinner, and consequently it was spread about the town that I was ruined. The effect upon shares that I held was instantaneous and tremendous. (ii, Ch. 74, p. 224.)

In depicting Melmotte's public reputation as the 'topping Croesus of the day', Trollope adopts the well-tried methods of Dickens and other earlier novelists. The expected hyperbole characterizes the general depiction of his business pursuits. 'He had concerns;—concerns so great that the payment of ten or twenty thousand pounds upon any trifle was the same thing to him . . . (i, Ch. 2, p. 19). He is said to have engaged in massive public works abroad, and to have victualled and equipped foreign armies (i, 4, 30–1), and he has power to influence the money-market whichever way he pleases. Inevitably he is 'fêted by aristocracy and politicians' (i, 4, p. 31), and in turn fêtes them in his magnificent town house with its usual trappings of luxury: the furniture, the carriages, the powdered and liveried footmen.

[6] Book, chapter, and page references are to the World's Classics edition of *The Way We Live Now* (Oxford, 2 vols. in one (1951), reprinted 1968).

[7] i, Ch. 40, pp. 379–80. The insincerity of Melmotte's words here and in the succeeding quotations, does not invalidate the point he makes. Credit and credibility are largely synonymous terms in business: 'credit' arises from the credibility of securities offered. Melmotte lives on various forms of credit in a credit-based ambience. His security, like that of Sadleir or Merdle, is merely 'the bubble reputation'.

Trollope does however, introduce into his novel a relationship that, in the context of the Victorian novel, is original. Early writers depicted the fraud and his Mammon-worshipping dupes; Trollope makes it clear that most of those involved with Melmotte know that he is probably a swindler. Rumours circulate that he was regarded in Paris 'as the most gigantic swindler that had ever lived' (i, Ch. 4, p. 31), and that he had been expelled from Vienna. He was thought to be utterly ruthless and treacherous, 'fed with the blood of widows and children' (i, Ch. 8, p. 68). He was 'thought by many to have been built upon sand' (i, Ch. 10, p. 84). Lord Nidderdale, initially at least, regarded him as 'a wretched old reprobate' (i, Ch. 22, p. 210), and to Dolly Longestaffe he was 'the biggest rogue out' (i, Ch. 28, p. 268).

Some of these rumours and views are true, others are not; their importance lies in the fact that the aristocratic figures in the novel are themselves so morally turgid that they are content to use a man whom they believe to be a fraud to enhance their own material prosperity, provided that he still retains his dishonest gains, and can make more. Roger Carbury, keeper of the standards of the older England, condemns Melmotte as 'a hollow, vulgar fraud from beginning to end' (ii, Ch. 55, p. 44); but it is the cultivators of Melmotte who receive his most powerful condemnation. 'These leaders of the fashion know,—at any rate they believe, that he is what he is because he has been a swindler greater than other swindlers . . . Men reconcile themselves to swindling . . . dishonesty of itself is no longer odious to them . . .' (ii, Ch. 55, pp. 44–5.)[8]

Influencing both Melmotte and his aristocratic colleagues is what Trollope saw as the 'American business ethic' of his time. A slick, crude ruthlessness and opportunism had afflicted society, epitomized in this novel by the promoter Hamilton K. Fisker, who, intoxicated with the concept of the 'big deal', sees in Melmotte a kindred spirit (i, Ch. 9, p. 76). He, like Fisker, will not be concerned with the viability of the Mexican railway, provided that its advance paper produces a profit. 'The object of Fisker, Montague and Montague was not to make a railway to Vera Cruz but to float a company.' (i, Ch. 9, p. 77.)[9]

[8] The lawyer Squercum, for instance, is delighted with his exposure of Melmotte mainly because the financier is a really grand villain: 'He regarded Melmotte as a grand swindler; perhaps the grandest that the world had ever known' (ii, Ch. 81, p. 290).

[9] Trollope accurately explains how it was possible to make a profit from company flotation in i, Ch. 9, p. 79.

The American view is developed in detail by Mrs Hurtle who, in her words to Paul Montague, contrives to give an appalling picture not only of her own ethics, but of those obtaining in English society. She admires Melmotte for his 'power', and 'grandeur'; he 'rises above honesty'; he recognizes that 'wealth is power, and that power is good'; the stronger a man is, even if he be unscrupulous, the nobler he must be (i, Ch. 26, p. 246). As the novel progresses, one realizes that Mrs Hurtle's words condemn her society more than they do Melmotte. He is by no means the abstract philosopher of power that she would have him be, but rather a man unable to control or even understand the forces that compel him to act as he does; he is also a particularly shabby and pathetic fraud.

It is as well to state here that no one is much inconvenienced by Melmotte's fall, except honest speculators like Mr Brehgert; those who, like Mr Longestaffe, have personal claims on his estate, are paid almost in full. Thus they learn nothing from their involvement with Melmotte, and cannot be considered his victims. He and they were conspirators in a new, amoral, and cynical society, in which, as Michael Sadleir put it, 'success was wealth, and wealth was God'.[10]

What was Melmotte's business? Two references in the novel suggest that he is a loan-jobber of the international type, 'arranging the prices of money and funds for the New York, Paris and London Exchanges' (i, Ch. 23, p. 217), and negotiating a 'Moldavian loan' (i, Ch. 40, p. 377). These are, however, examples of traditional decoration: Trollope does not insist on them, and they are lost in the vastness of the novel. Melmotte's principal activity is that of company promoter, and we learn that he is a director of three dozen companies.[11] His firm is called 'Melmotte & Co.', but, as Trollope drily remarks, 'Of whom the Co. was composed no-one knew . . . he had never burdened himself with a partner in the usual sense of the term.' (i, Ch. 9, p. 81.) It is thus in effect a private firm, with its proprietor the sole 'partner'; Melmotte is free to use it as he wishes.

It is Melmotte's reputation as a 'big dealer', with a history of bold if disreputable transactions, that draws him into that great project, the 'South Central Pacific and Mexican Railway'. Through Fisker's

[10] Sadleir, p. 398.

[11] Melmotte operates from Abchurch Lane, so that he keeps company with Rothschild, Sidonia, and Osalez. His place of business, in a 'small corner house', with 'narrow and crooked' steps, and 'small and irregular' rooms, shares their reassuring dinginess (i, Ch. 9, p. 81).

influence he becomes chairman of 'the British branch of the Company', which he is able to manipulate and plunder at will. The several descriptions of Board meetings, so reminiscent of those 'polite little fictions' held by Tigg Montague thirty years earlier, signal the direction in which the novel's action will progress. We learn that 'everybody would agree to everything, somebody would sign something, and the "Board" for that day would be over.' (i, Ch. 22, p. 205.) The records of previous meetings, read by Miles Grendall, were actually written by 'a satellite of Melmotte's from Abchurch Lane' (i, Ch. 37, p. 342).[12]

Trollope introduces into his novel a new dimension of business hypocrisy that suits well with the climate of the sixties and seventies, a pretence that private gain is sought for the public good. Everyone connected with the railway company is there to make his fortune by 'manufacturing the shares thus to be sold'. 'But now . . . they talked of humanity at large and of the coming harmony of nations.' (i, Ch. 10, pp. 89–90.) Melmotte himself declares that his railway will 'afford relief to the oppressed nationalities of the over-populated old countries', and enable 'young nations to earn plentiful bread' (i, 44, p. 412).[13] These words are applauded by the aristocrats and gentry who attend the Board and whom Trollope sees as betrayers of their class and their country. Little wonder that Roger Carbury declares of England: 'It's going to the dogs, I think;—about as fast as it can go.' (ii, 55, p. 45.)

In establishing and developing the role of Melmotte, Trollope was primarily concerned to condemn those who professed to believe in him, and who cultivated him because 'it is always wise to have great wealth on one's side' (i, 4, p. 30). Dickens had had the same general intention in *Little Dorrit*, but had been careful, while showing Merdle as in part the victim of the society he worshipped, not to arouse our sympathy for him. Trollope, however, becomes more involved with his creation,

[12] Secure in the knowledge that his 'Board' of tame and ignorant aristocrats will not question his judgement, Melmotte proceeds to manipulate the company's paper to his personal advantage, and, where necessary, pays dividends out of capital. See, for instance, his attempts to disarm Paul Montague in this manner (i, Ch. 10, p. 85; i, Ch. 22, p. 206), and his payment of Sir Felix Carbury in virtually worthless railway scrip (ii, Ch. 67, pp. 155–6).

[13] Melmotte usually adopts a cheerful, confident tone in his boardroom. (See, for example, i, Ch. 37, p. 345.) However, at the farewell dinner for Fisker at the Beargarden, he seems to be off his guard, and Trollope depicts him with the bland mask of confidence for once removed. He is nervous, his speech halting, and his gaze averted from his companions (i, Ch. 10, p. 89). This is very like Merdle; but the mood does not occur again in the novel.

and his disgust with society at large makes him endow Melmotte with some qualities that raise him above the sycophants who surround him. Thus Trollope's stance changes from that of external observer of a public image to that of a confidant, who is able to reveal to the reader what might be termed Melmotte's 'inner character'.

At first, in Book I, Melmotte is depicted as the 'horrid, big rich scoundrel' (i, 23, p. 220), and his parvenu qualities are detailed. Like Merdle, he is proud to be considered part of Society, even though, unlike Merdle, he bullies it; a more forceful person than Dickens's creation, he is nonetheless afflicted with the same disease. In Book II, however, Trollope involves Melmotte in a duel between two conflicting emotions, fear and fortitude, and it is in the resolution of this personal conflict that the interest of Melmotte as a character now lies. When the City notables fail to attend his banquet for the Emperor of China, he is 'bewildered and dismayed': 'He was now striving to trust to his arrogance, and declaring that nothing should cow him. And then again he was so cowed that he was ready to creep to anyone for assistance.' (II, Ch. 59, p. 85.)

Fear and basic insecurity are aspects of Melmotte new to the reader, and these are developed as the story progresses. Melmotte is not fully master of his schemes, and fears that circumstances might carry him into 'deeper waters than he intended to enter'; this knowledge is a burden that 'might crush him at any time' (ii, Ch. 62, p. 104). In similar circumstances Merdle had shown fear, which had been succeeded by a deadly calm prior to his death. Melmotte reacts differently, striving to meet his fears with fortitude and stoicism: Trollope now admits that 'there was a certain manliness about him', though conceding that 'fraud and dishonesty had been the very principle of his life' (ii, Ch. 81, p. 295). When Trollope begins to look inside his financier, he produces this interesting apologia, a combination of moral courage and moral imbecility.

Melmotte resolves never to run away from his difficulties, but rather to 'bide his ground . . . with courage' (ii, Ch. 62, p. 105); in the face of arrest he would 'go through it, always armed, without a sign of shrinking' (ii, Ch. 63, p. 119). These resolutions compel a grudging admiration, and this is enhanced by his declaration to his wife that he had not had it 'very soft' all his life, and that 'whatever there is to be borne, I suppose it is I must bear it' (ii, Ch. 81, p. 299).

The human dimension is further developed by the loyalty exhibited towards Melmotte by his subordinates. Herr Croll makes no attempt to expose his master, though unwittingly involved in forgery: 'He had not behaved unkindly. He had merely remarked that the forgery of his own name half-a-dozen times over was a "strong order".' (ii, Ch. 82, p. 302.) Mr Merdle's butler, on hearing of his master's death, wishes to give a month's notice;[14] Melmotte's butler, breaking the news of his death to Lord Nidderdale, declares, 'I think you'll be sorry to hear it', covers his face and bursts into tears. (ii, Ch. 85, p. 334.)

Closely allied to this revelation of Melmotte's 'inner character' are a series of passages throughout the novel which emphasize Melmotte's growing lack of prudence, and his inability to control his own schemes. The game that he had intended to play, we learn 'had become thus high of its own accord', concerning which Trollope remarks that 'a man cannot always restrain his own doings . . . He had contemplated great things; but the things which he was achieving were beyond his contemplation.' (i, Ch. 35, p. 323.)

As the novel develops, Trollope reveals how Melmotte's transactions become too much for him to contain, and that their complex progress is due to the dangerous operations of credit, and to Melmotte's increasingly arrogant belief in his own infallibility. His 'arrogance in the midst of his inflated glory was overcoming him' (i, Ch. 45, p. 428) '. . . there had grown upon the man . . . an arrogance, a self-confidence inspired by the worship of other men, which clouded his intellect . . .' (ii, Ch. 53, p. 20).

He becomes 'deficient in prudence' (ii, Ch. 54, p. 34), and is so dangerously deluded that 'he came almost to believe in himself' (ii, Ch. 56, p. 57). Thus 'his ambition got the better of his prudence' (ii, Ch. 62, p. 105), and he embarks upon the series of acts that leads to fraud and death.

Melmotte is destroyed by lack of credit, by the inability to raise immediate cash when it is required—a failure that he shares with John Sadleir. It is wrong to imagine that he was 'ruined': he died not really through lack of money, but through lack of time. Ironically he falls victim to his own complex schemes—'he did not really know what he owned, or what he owed' (ii, Ch. 73, p. 219)—which had placed his only source of immediate credit in the hands of the absconding Cohenlupe, who proves to be a greater, because a wiser, rogue: '. . . he had allowed himself to become hampered by the want of comparatively small sums of ready money, and in seeking relief had rushed from one

[14] *Little Dorrit*, ii, Ch. 25.

danger to another, till at last the waters around him had become too deep even for him, and had overwhelmed him.' (ii, Ch. 86.)

The crowning irony of Melmotte's career is that his anxiety to fit into 'Society' causes him to purchase Pickering Park from Mr Longestaffe.[15] 'His breath was taken for money', and Longestaffe is content to wait for payment; in the interim Melmotte secretly mortgages the property before he is able to pay for it. Thus the Great Financier will founder over what is essentially a private, domestic transaction.

The idle, stupid, dissipated son, Dolly Longestaffe, proves to be the disreputable Nemesis of this story. Ignorant and vapid, these very limitations permit him to see beyond the sophistication of his elders. He wants to know 'when Melmotte is going to pay up this money': 'I don't know why Mr. Melmotte is to be different from anybody else . . . When I buy a thing and don't pay for it, it is because I haven't got the tin, and I suppose it's about the same with him.' (ii, Ch. 58.)

Thus it is that the lawyer Squercum is set by Dolly to ferret out the truth. Squercum discovers that Melmotte and Cohenlupe are engaged in share-rigging, and that Melmotte is involved in a particularly shady property deal. What impresses the reader is the unglamorous nature of these transactions. The 'topping Croesus of the day' mortgages another man's house, and destroys himself by indulging in practices that are shabby, furtive, and unimaginative.

As rumours spread, confidence falls away; and Melmotte is seen at work on his pathetic and despicable criminal forgeries, tracing signatures, and destroying incriminating documents by fire and even by swallowing. When all his subterfuges fail, and arrest over the Longestaffe property is inevitable, Melmotte makes his last terrible appearance in Parliament, intoxicated, and incoherent; then returns home and kills himself: '. . . at nine o'clock on the following morning the maid-servant found him dead upon the floor. Drunk as he had been . . . still he was able to deliver himself from the indignities and

[15] 'The family of Longestaffe are Trollope's most merciless comment on the degeneration of the squirearchy. Ambitious and scheming, they are also meanly inefficient. They put an honoured name in pawn to city charlatans . . . At home they snarl and scrape and plot for pin-money and husbands; abroad they haughtily maintain at least the attitudes of a betrayed nobility.' (Sadleir, p. 400.)

penalties to which the law might have subjected him by a dose of prussic acid.' (ii, Ch. 83.)[16]

Few would question the fact that Mr Merdle was an Englishman, but throughout *The Way We Live Now* Trollope ascribes a distinct foreignness to Melmotte.[17] He had first been known in England as M. Melmotte, and he spoke English with an accent, suggesting 'at least a long expatriation'. However, he declared that he was a true Englishman, born in England. Marie Melmotte, his daughter, had been born in 'the German portion of New York', and as a child had moved with her father to Hamburg, where they lived in dire poverty.[18] In Frankfurt, home of the Rothschilds, the family had become Jews, but reverted to Christianity in Paris. From there they had removed first to Brighton, then to London. After his death, a 'general opinion' is held that Melmotte's real name was Melmody, and that he was the son of a 'noted coiner in New York'. Wherever he was born—Ireland, England, or the New York ghetto—Melmotte acknowledged in secret that he was 'a boy out of the gutter'.

It was shown in the previous chapter how Dickens had had the 'general idea' for Merdle in his mind for many months before the death of Sadleir provided him with a real-life model. It is possible that Trollope, in creating Melmotte, modelled him at least in part on a contemporary financier whom he must have heartily disliked, Baron Albert Grant.[19] Grant, like Melmotte, was of 'obscure origin'—son of a German Jew, W. Gottheimer, who ran a foreign fancy-goods business

[16] Melmotte's death by prussic acid parallels that of John Sadleir, but there is no need to see a connection here. This poison seems to have been freely available at the time, and suicides effected by this substance and by spirits of salt (hydrochloric acid) were frequent.

[17] Attempts have been made to connect the name Merdle with a gross French vulgarism. (See, for example, Barbara Hardy, *The Moral Art of Dickens* (1970), p. 20.) One is very much at liberty to doubt this interpretation, and at the same time prompted to suggest that 'Melmotte' may signify 'honey-tongued', describing a man whose honeyed words persuade others to believe in him.

[18] He asserts (at i, Ch. 4, p. 30) that he was born in England. Trollope tells us later in the novel that 'he had once been imprisoned for fraud at Hamburg, and had come out of gaol a pauper' (ii, Ch. 81, p. 298).

[19] Escott, p. 297, states: 'Melmotte himself carries about him a certain suggestion of Baron Albert Grant in the past, and of Whitaker Wright in the days that were yet to come . . . Trollope's Melmotte was an exaggerated type of the French, German and American adventurers who . . . gorge like vultures on the country.' The financier Whitaker Wright defrauded shareholders of the London and Globe Finance Corporation, and two other enterprises, of £5,500,000. Extradited from the USA in 1903, he was sentenced in the following year to seven years' penal servitude; he committed suicide on the same day.

in London. Born in Dublin in 1830, he early changed his name to Grant. He was the pioneer in England of mammoth company promoting, obtaining vast sums of capital from small investors: 'All sorts of kind individuals were at his elbow, ready to supply him with the means of meeting the demand, and he was tempted into embarking upon schemes without proper investigation.'[20]

The names of Grant's companies reveal how closely he followed the practices of the French public works projectors. Among his companies were the Belgian Public Works, the Central Uruguay Railway, the Crédit Foncier et Mobilier of England, and the Imperial Bank of China. These grandiose schemes were largely worthless, and many people lost their money, although Grant made a large fortune.

In 1865 he became MP for Kidderminster, being re-elected in 1874; he was created baron by King Victor Emmanuel in 1863 for services rendered in connection with building in Milan. At home, he purchased and laid out Leicester Square, presenting it to the city; he also purchased pictures for the National Portrait Gallery. After 1875 numerous aggrieved investors began actions for restitution; in June 1877 eighty-nine such actions were pending. A year earlier, Grant had been acquitted on a charge of fraud. Grant's wealth rapidly dwindled away, and his affairs went into liquidation in February 1879. Grant lived for another twenty years, but even at his death, on 30 August 1899, a receiving order had been made against him at the London Bankruptcy Court.[21]

Some revealing remarks about Baron Grant are to be found in S. Baring-Gould's *Early Reminiscences* (1923), pages 52–5. Baring-Gould, in these pages, gives a sneering, hostile account of the adoption by European Jews of distinguished gentile surnames, and then proceeds to give the following rather inaccurate account of Grant:

> A German Jew of the name of Gottheimer came to England as a company promoter, and assumed the name of Albert Grant, Grant being the family name of the Earls of Seaforth, the heads of the Clan of Grant . . . Heaven and the Court that ennobled him only know how and for what he obtained his title as Baron Grant. But the epigram concerning him circulated freely:
>
> 'King may a title give,
> Honour they can't,
> Title without honour
> Is a *barren grant*'.

[20] T. Seccomb, 'Grant, Baron Albert', in *DNB* vol. xxii, from which this brief account is taken.

[21] See his obituary in *The Times*, 31 Aug. 1899, p. 4.

The fellow died comparatively poor in 1899, owing to a series of actions in the bankruptcy court.

The contemptuous tone adopted here, and the unpleasant anti-semitic quality of the pages cited, would have appealed to Trollope and his contemporaries,[22] and the apparent Jewishness of Melmotte, initially at least, was no doubt meant to provoke that particular contemporary response. Melmotte, like Grant, may be seen as one of the 'wrong Jews' who were said to be vitiating society in the seventies: 'The international financial adventurer had settled on London in his swarms . . . the wrong Jews came ever more blandly to the right houses . . .'[23] This aspect of Melmotte partly distinguishes him from Mr Merdle, and is at least worth noting when one is seeking similarities and differences in these two literary creations.[24]

While denying the pessimistic gloom of Carlyle and Ruskin, Trollope feared that 'men and women will be taught to feel that dishonesty, if it can become splendid, will cease to be abominable',[25] and this view caused him to go 'beyond the iniquities of the great speculator who robs everybody', and to castigate the folly and mean shoddiness of a vitiated society. While admitting that the accusations are exaggerated, and the 'implied vices coloured', Trollope declared that, 'I by no means look upon the book as one of my failures; nor was it taken as a failure by the public or the press.'[26]

The reviewer of his novel in the *Graphic* would have demurred:

Mr. Trollope's new story is disappointing . . . because of the repulsively disagreeable society to which it introduces us . . . We are made, as it were, to breathe an atmosphere of sordidness and knavery till it becomes oppressive and tiring . . . As a 'study' of human baseness and mean depravity 'The Way We Live Now' must be pronounced successful; but it is not pleasant reading, and few, we should think, would care to turn to it a second time.[27]

The *Graphic* found Trollope's novel 'oppressive and tiring', and one reason for this response is that the theme of Mammon, and the castigation of an acquisitive society that accompanied it, was becoming

[22] See the parody of Disraeli's Sidonia in Trollope's *Barchester Towers* (1857), Ch. 9, examined in Ch. 5 above.

[23] Sadleir, p. 398.

[24] There is an interesting and sensitive discussion of Melmotte's Jewishness in E. Rosenberg's *From Shylock to Svengali* (1961), Ch. 6: 'The Jew as Parasite'.

[25] A. Trollope, *Autobiography* (1883), World's Classics edn. (1968), p. 304.

[26] ibid., p. 306.

[27] The *Graphic*, 17 Jul. 1875, p. 62.

worked out. This conflict between the world of letters and the world of the City, rooted partly in landed snobbery, partly in righteous reaction to some of the excesses of the new industrial society, had been fought, in novel and essay, at least since the twenties. But by 1875 new attitudes were beginning to prevail. Society, with its Queen proclaimed Empress of India, sought to look beyond its own shores to the exciting concept of Empire. The reader was looking for different experiences, and a new generation of writers, Walter Pater and Henry James, R. L. Stevenson and Oscar Wilde, turned their talents to the perfection of style, the polished phrase, the subtle, the mystic, the esoteric.

By the nineties, the battle of City magnate against the old landed interest had largely been dissipated into an uneasy and convenient truce, resulting in marriages and alliances of the old and the new. The inner conflict, of course, still raged; certainly one of the most devastating dramatic re-creations will be found in Galsworthy's *The Skin Game* (1920). But Trollope registered his protest at a time when both critics and public were beginning to seek new interests, and to tire of the reiteration of old themes.

Mr Bradford A. Booth asserts that 'to trace Melmotte through Merdle is an idle exercise',[28] and of course he is right: Melmotte is a more highly developed character than is Merdle; more domineering, flamboyant, and, curiously, more vulnerable and pathetic. However, the general similarities of their careers and fates arise from their both being products of the rapid development of credit finance, originating in France and spreading quickly to England, where its operations were enhanced by the protections afforded through limited liability. Dickens and Trollope, writing twenty years apart, were responding to different stages of the same commercial stimulus. An acquaintance with the financial history of the nineteenth century can, in many instances, throw light not only upon the motives that impelled their authors to create them.

[28] Booth, 'Trollope and *Little Dorrit*', p. 240.

8
George Hudson and Railway Promotion

Of all the dramatic changes wrought by the industrial revolution on the daily life of nineteenth-century people, none was so far-reaching in its effects as the advent and progress of steam locomotion. Within a period of some fifteen years the country seemed to shrink physically as principal cities and towns, villages and even hamlets, were linked one to the other by the permanent way. The year 1825 was noteworthy not only for the Company Mania, but for the opening in October of the Stockton and Darlington Railway. This event ushered in the 'Railway Movement', as it was called, a momentous stride forward in the history of transport, and an opportunity for new, vast commercial enterprise. In the precarious world of joint-stock undertakings, with which the reader is now familiar, new consortia and personalities arose, whose vast schemes and colossal dreams, often realized in fact, put earlier enterprises in the shade.

Yet, when one reads those novels of the period that feature railway material, one is surprised how little is made of the commercial aspects of railway promotion. In the main, novelists and their public were more impressed by the railways' potential for physical destruction than they were by the possibilities offered by the frequently dramatic careers of the railway promoters. There are many descriptions of violent death and fearful accidents, and these can be traced back to the truly horrifying accident that occurred almost coincidentally with the foundation of the railway system.

This accident happened at the formal inauguration of the Liverpool and Manchester Railway, on 15 September 1830. At eleven o'clock in the morning, a grand procession of eight locomotives and a train of twenty-nine carriages left Liverpool for Manchester. There were six hundred passengers, including the Duke of Wellington, Sir Robert Peel, and William Huskisson, the distinguished politician and Member of Parliament for Liverpool. At Parkside the engines stopped for water, and, ignoring instructions, many of the passengers left the carriages, and stood on the permanent way, which consisted of two lines of rails. Huskisson went to speak to the Duke of Wellington just as several engines were seen approaching. There was a hasty scramble back into the carriages, but Huskisson, who was clumsy and hesitant by nature, lost his balance and fell back upon the rails in front of the 'Dart'. He was fearfully injured, and lingered for nine hours before he died. He was buried in Liverpool on 24 September.[1]

The grave sense of waste and loss, and the terrible irony of such a tireless advocate of this railway being killed at its opening, still strike one today with a sense of shock, and it is not surprising that this event would provide an exemplar for many subsequent fictional depictions of railway slaughter and carnage.[2] Probably the most widely known use by a novelist of this demonic aspect of the new mode of transit is the retribution meted out to the evil Carker in *Dombey and Son* (1846), where, falling beneath an approaching train, the wretched man is 'struck limb from limb' in a celebration of lurid and tasteless language.[3] Many other examples of this kind of writing can be found in novels of the period, where accidents and derailments provided either dramatic incident, or a means of removing a villain from the plot. In no case are we presented with a truly coherent account of railway life, be it in the board-room or on the foot-plate.[4]

Some novels show what may be termed social concern at the changes wrought by the railways, but always as peripheral to the interest of whatever 'plot' is being unfolded. The intrusion upon private rights worried Robert Bell, who, in *The Ladder of Gold* (1850) declared that 'physical obstacles and private rights were straws under the chariot

[1] This brief account is taken from the article on Huskisson by J. A. Hamilton (Lord Sumner) in *DNB*.

[2] Regular accidents and derailments characterized the first decades of rail transport. In 1841, for instance, there were 29 accidents involving collision and derailment, resulting in 24 deaths and 71 injured. In a single week of August 1845 there were 17 accidents on 9 lines, resulting in many injuries. On some lines in the forties signalling equipment was entirely lacking, and not all companies fixed gradient-boards adjacent to the tracks. See J. W. Dodds, *The Age of Paradox* (1953), p. 217; H. Grote Lewin, *The Railway Mania and its Aftermath* (1936), p. 111.

[3] The passage will be found in *Dombey and Son*, Ch. 55, which appeared in 1847. Robin Atthill, in his article 'Dickens and the Railway', *English*, No. 13 (1961), pp. 130–5, suggests that the importance of the railway throughout *Dombey and Son* is in part due to the events of the 'Railway Mania'.

[4] See, for instance, D. Costello, *The Millionaire of Mincing Lane*, in *Bentley's Miscellany*, vol. 41 (1857), p. 647, where there is much blood and scalding steam; W. G. Wills, *Notice to Quit* (1861), vol. ii, p. 147 f., where the death of Huskisson is re-enacted; H. J. Byron, 'Paid in Full' in *Temple Bar*, vol. 13 (1865), p. 107, where the wicked lawyer is devoured by the 'glowing demon'. A variety of fearful railway deaths and mutilations is offered in J. B. Harwood's 'Lord Ulswater', in *Chambers's Journal* (1867), p. 474.

wheels of the Fire-King . . . sweltering trains were to penetrate solitudes hitherto sacred to the ruins of antiquity . . .'.[5] The anonymous author of *Yesterday* (1859) describes the coming of the 'navigators' to a sleepy village, where they startle 'the holy quiet of a Sunday afternoon by hoarse songs of ruffian mirth, or hoarser bellowings of ruffian fury'.[6] The younger Albany Fonblanque, in *A Tangled Skein* (1862), lamented that the railways had blurred the distinction between town and country. F. R. Chichester, in *Masters and Workmen* (1851), saw the exposure of slums to the light of day by railway excavation as a spur to social progress: 'These plague spots of our great commercial towns are no longer hidden by plaster palaces and flaunting gin-shops . . . but . . . arouse the passer-by to a sense of the duty and necessity of endeavouring to improve the moral and physical condition of those whom destiny has condemned to inhabit such abodes . . .'[7]

One or two novels, and a fair output of ephemeral, journalistic tales and articles, do reflect the financial and commercial aspects of railway promotion, and these will be examined at relevant points later in the chapter. As literary works they are interesting in that some provide accurately observed reflections of the two 'Railway Manias' and their accompanying commercial crises, while others use these events to create largely misleading stereotypes of their dominant personalities.

Of all the great promoters connected with railway expansion, none absorbed the attention of the professional writer more than did George Hudson, the 'Railway King'. Hudson was born in 1800, and at the age of fifteen was apprenticed to a firm of drapers in York. With an excellent head for business, he quickly rose to be a partner in the house, and by 1827 he was a man of considerable substance. In that year he was left a legacy of £30,000 by a distant relative, and with an unerring instinct about the potential of the new railways, he invested the whole of this sum in the shares of the recently floated North

[5] Op. cit., i, pp. 273 f. Mrs Gore also regretted the passing of the old character of village life. In her novel *Progress and Prejudice* (1854), she regretted that 'We are grown more locomotive' (i, pp. 12 f.).

[6] Op. cit., pp. 260 f. The early navvies were indeed ignorant and brutalized, and as a class they spread terror through the countryside. In Chadwick's Report of 1846 they were described as drunken and dissolute. (See J. W. Dodds, op. cit., p. 245.) A more humorous judgement on the 'navigators' and their role is in E. D. Cook's 'The Trial of the Tredgolds' (1864), in *Temple Bar*, viii, p. 408. A stern condemnation of them, 'brutalised and degraded, retaining nothing of humanity but its form', is in S. W. Fullom's *The Great Highway* (1854) i, pp. 264 f.

[7] Op. cit., i, pp. 16 f.

Midland Railway Company. Within a decade he was to become the chief architect of the nation's railway system, and was to earn for himself the not ill-deserved sobriquet of 'Railway King'.

In 1833, three years after the death of Huskisson, George Hudson became leader of the Tory interest in York, and in the same year established the York Banking Company, a joint-stock enterprise which, unlike many country banks, 'did *not* ruin its shareholders, but on the contrary (and chiefly through Mr. Hudson's excellent management), withstood all shocks to its credit, and became a "paying" concern.'[8] Two years later he was elected a town councillor, and in 1836 he became an alderman. This was the year in which the 'Little Railway Mania' began.

The 'Little Mania' was not an isolated phenomenon, but part only of a further episode in the ten-year cycles of boom and depression which seemed to characterize the commercial life of the century. In 1836 the country was enjoying a period of prosperity, and business in most fields was buoyant. Such conditions always produced low rates of interest, so that money was seen to be plentiful. In 1834, £10,600,000 of Government Stock had been reduced from 4 to 3½ per cent, and there was much 'blind capital' seeking returns of 5 per cent or more. Investors were thus encouraged to put their money in railway stock, which promised a high yield on investment: 'The press supported the mania; the Government sanctioned it; the people paid for it. Railways were at once a fashion and a frenzy.'[9] During 1836 and 1837, thirty-nine railway Bills received assent.

This sudden rush for railway paper triggered off a general mania, and many of the joint-stock banks, which numbered forty-five by 1836, encouraged the frenzy by offering easy accommodation. There was a sharp wave of speculation and over-trading, with a rash of railway projects, and companies for distilleries, insurance, cemeteries, newspapers, sperm oil, cotton twist, and other commodities.

The pattern of events now echoed that of 1826. The Bank became alarmed, and communicated its alarm to the other banking houses.

[8] *Fraser's Magazine*, Aug. 1847, p. 218.

[9] J. A. Francis, *A History of the English Railway* (1851) i, p. 290. Some of these investments paid off. Among important railways floated in these two years were the Midland Counties, North Midland, Great North of England, and London and Brighton. (See Clapham, op. cit., i, p. 388.) In all periods of heightened speculation during the century, many creditable and valuable institutions were established. Not all speculative enterprises were 'bubbles'.

Discounts were suddenly diminished, and money became 'tight'. Trade ground to a halt, and 'bubble' enterprises, and some of the 'paper' railway companies, failed. This sharp and sudden curtailment of trade slowed down railway development until the 1840s, but overall its effects were neither lasting nor particularly severe. Those railway companies that were properly capitalized continued to consolidate and expand.[10]

An interesting reflection of the 'Little Railway Mania' is found in one of Dickens's minor works, *The Mudfog Papers*, where, in 'The First Meeting of the Mudfog Association for the Advancement of Everything', published in *Bentley's Miscellany* in October 1837, the author links the speakers at Mudfog with the activities of speculators and others at the time. Dickens cleverly reveals the frantic activity of the jobbers in share trafficking, and the unsoundness of many of the 'paper' lines floated during the period. 'Mr. Jobba', for instance, produced 'a forcing machine . . . for bringing joint-stock railway shares prematurely to a premium', a device which indirectly castigates directors who 'puffed' the paper of new companies until they could part with it at a profit to the hapless shareholders. 'A Member . . . wished to know whether it was not liable to accidental derangement? Mr. Jobba said that the whole machine was undoubtedly liable to be blown up, but that was the only objection to it.' Here we see Dickens the journalist reflecting what was in effect a limited contemporary preoccupation.[11]

In the year that the 'Little Mania' came to an end, George Hudson became Lord Mayor of York. In the early thirties the city had evinced a conservative mistrust of railways, but activities in the rival town of Leeds began to effect a change of attitude. Railways were thrusting upward from London to the North, and Leeds had begun a line running eastward to Selby, to form a junction with the Hull and Selby

[10] The account of the 1836–7 crisis given here is derived in the main from Morier Evans, *History of the Commercial Crisis 1857–1858* (1859), pp. 16–20, and from Clapham, op. cit., i, pp. 388–9. From 1825 to the close of 1835, 54 railway Acts were passed, including that for the establishment of the London and Birmingham line, 112 miles long, with a capital of £5,500,000. This line, opened in 1838, accounted for nearly half of the 500 miles of public locomotive railway then operating (Clapham, i, pp. 387–8).

[11] Professor G. A. Chaudhury has explored in depth what he has called the 'facetious topicality' of *The Mudfog Papers* in *The Dickensian*, May 1973, p. 104. He points out that the British Association for the Advancement of Science, which had been founded at York in 1831, was thought to vulgarize science, and that this laid it open to satirical treatment. A parody 'report' had appeared in Cruikshank's *Comic Almanack* for 1835, and the opening number of *The Pickwick Papers* (April 1836) satirized the style of the papers read at the Association's meetings (Chaudhury, loc. cit., p. 105).

line, thus securing direct rail linkage to London. Access to the capital had constantly exercised Hudson's mind, and this action from Leeds determined him to project a line from York which would join with the existing North Midland Railway Company. Largely through Hudson's vigorous activity, a capital of £446,666 was raised, and an Act of 1837 incorporated Hudson's proposed line with that already existing to form the York and North Midland Railway Company. The earlier railway had not been too profitable an enterprise, and Hudson had offered to re-establish it on a sounder and more economic footing. The line from York was opened in May, 1839, and the first trains ran in the following year.

From this time onward, Hudson was established as the principal railway promoter in the land. His brilliant planning, his bold and fearless approach, and his remarkable ability to forecast the potential profitability of routes, ensured for him the direction of many lines. It was Hudson who developed the principle of incorporation and amalgamation, adding financial and administrative strength to otherwise poorly-subscribed and fragmented companies. The popular title of 'Railway King' was a tribute to a man of forceful genius. Constantly concerned with efficiency, Hudson in 1842 established the railway clearing system, which came into use on two lines in January. Concerned with the smooth operation of through traffic, the clearing house in Seymour Street, Euston Square, adjusted the financial relations of the various companies by means of a number of departments, such as secretarial, merchandise, and mileage.

By 1842 money was cheap again, and the country was ready and willing to indulge in another round of investment. When a line via Newcastle to Edinburgh was proposed in June of that year, Hudson was appointed chairman. He subscribed five times as much as any other director, and personally guaranteed a return to shareholders of 6 per cent. At the same time he raised a capital of £5 million, and amalgamated three lines approaching Derby into the Midland Railway Company.

By 1844 Hudson controlled 1,016 miles of railway; all the companies in which he was engaged were successful. The investing public, enthralled by his very name, rushed to subscribe in any line connected with him, and, indeed, in any new railway company, whether soundly based or a mere fiction of jobbers. In the course of January 1845, sixteen new railway companies were registered. The share-market was stimulated to over-activity, and in April fifty-two companies came into being. Inevitably, the dramatic increase in the number of lines projected, and the availability of cheap capital, led to a rage for speculating in pig-iron, which was used in the manufacture of rails. By the time spring arrived, the country was in the grips of yet another speculative convulsion, the 'Great Railway Mania'.

As in 1837, novelists and other writers responded to the mania with various journalistic pieces. The pages of *Punch* abounded with material on the railway fever, and *Punch's Almanack* for 1846 is devoted entirely to railway matters, as is Cruikshank's *Table Book* for 1845. Of particular interest are Thackeray's contributions to *Punch*, 'The Diary of C. Jeames de la Pluche, Esq., with His Letters', which appeared on various dates throughout 1845 and 1846.[12]

James Plush, a footman, 'speculated in railroads', and his 'winnings' were £30,000. Immediately accepted into Society, and equipped with an instant Norman pedigree and name, he is soon director of thirty-three railways, and intends to stand, like Hudson, for Parliament 'on decidedly conservative principles'. A 'Letter from Jeames' follows, in which he explains how he founded his fortune on a £20 loan from a fellow servant: 'That I've one thirty thousand lb, *and praps more*, I don't deny. Ow much has the Kilossus of Railroads one, I should like to know, and what was his cappitle?' (p. 105.)

We further learn that Jeames christened his first pair of horses Hull and Selby, a grateful allusion to his successful investments in that company. A shareholders' revolution in 1845 had resulted in the leasing of the Hull and Selby to Hudson, who offered very generous terms,[13] so that Jeames was to be congratulated on his investment. His riding cob he 'very unaptly' called 'Dublin and Galway', which suggests that he (and Thackeray) recognized a non-starter: this line, more properly called the Irish Great Western, was mooted in 1845, but its Bill was rejected. Jeames called his valet 'Trent Vally', as he had made a profit from 'that exlent line'. Authorized in 1845, the Trent Valley was one of the most important to be established, providing a main line between Rugby and Stafford. Its completion provided the Manchester and Birmingham Company, who sponsored it, with a considerably shorter

[12] *Jeames's Diary* appeared in *Punch*, 2 Aug. 1845 to 7 Feb. 1846. It was reprinted in vol. ii of *Miscellanies: Prose and Verse* (1856). Page refs. in the text are to *Works* (Smith, Elder & Co.), 1879, vol. xv.

[13] H. Grote Lewin, *The Railway Mania and its Aftermath*, pp. 77–8.

THE STAGS. A DRAMA OF TO-DAY.

DRAMATIS PERSONÆ.

TOM STAG, *a Retired Thimblerigger.*
JIM STAG, *an Unfortunate Costermonger.*

(TOM *dictates to* JIM).

NAME IN FULL . . . . . . *"Victor Wellesley Delancey."*
RESIDENCE . . . . *"Stagglands, Bucks."*
PROFESSION . . . . *"Major-General, K.C.B., K.T.S., K.S.W."*
REFERENCE . . . . *"His Grace the Duke of Wellington.*
"Sir Robert Peel. Coutts and Co."

"That'll do. Now, MARY, a wafer: and, JIM, I don't mind standing a pint of alf and alf!"

4. Dishonest company promoters frequently claimed noble patronage for their 'bubble' enterprises.

and more direct line to London, and once again Jeames is to be congratulated on a shrewd investment.[14]

The article 'Jeames on Time Bargings' accurately reflects a particular activity of speculators at this time. A time bargain was a Stock Exchange contract for the sale of a certain amount of stock or number of shares on a future day at a fixed price rather than the actual market price on that future day. Such contracts were barely-concealed gambles on the movement of prices, and though many paid off, they were regarded as being commercially disreputable.

In Thackeray's tale, young Frederick Timmins, with an annual income of £200, is seduced into railway speculation by his fellow club members, who had made small fortunes in the market, and he is soon smitten by speculation fever. A month before the supposed date of Jeames's narrative, young Timmins had bought a hundred shares in the 'Grand Niger Junction, or Gold Coast and Timbuctoo (Provisional) Atmospheric Railway'.[15] These £20 shares could be had for 9*d.* paid (in itself a remarkable bargain!), so that Timmins parted with only £3.15*s.* He immediately sold the shares for £250. In this he showed himself to be a bear, speculating for a fall, because he is very annoyed when, in a couple of days' time, the price of Atmospherics rose to £5.

At this point, he is persuaded by a secret foe, Colonel Claw, to contract a time bargain. The Colonel offers a very enticing 'tip' to the young speculator:

'My dear fellow, the shares will be at 15 next week. Will you give me your solemn word of honour not to breathe a word to mortal man what I am going to tell you?'
'Honour bright,' says Fred.
'HUDSON HAS JOINED THE LINE.'[16]

[14] Morier Evans tells us that the Prime Minister himself cut the first sod of the Trent Valley Railway, and had spoken highly of the company (Evans, op. cit., p. 13). See also Grote Lewin, pp. 34–8. Jeames further mentions his connection with the North British, London and Birmingham, Manchester and Leeds, Bristol and Exeter, and Eastern Counties lines. All these were engaged on schemes of consolidation and expansion during 1845.

[15] Clegg and Samuda's atmospheric system of propulsion was a serious alternative to locomotive traction. Brunel favoured the system, and it was used on parts of the London to Croydon line. However, the insuperable difficulty of sealing the longitudinal valve after the passage of the piston carriage-arm led to the abandonment of the system in the late forties.

[16] 'he that in 1845–6, can command the nod of a Hudson may be supposed to reckon upon unlimited premiums in all kinds of lines . . .' (Morier Evans, p. 11).

LORD BROUGHAM'S RAILWAY NIGHTMARE.

5. The old jurist is disturbed by the volume of railway paper coming before Dalhousie's Advisory Board during the Mania years.

Timmins rushes into the City, and tells his broker to buy one thousand Great Africans, which is done for him at four and seven-eighths. Alas! When the thirtieth of the month arrives, the shares are priced at one quarter for 9*d.* paid. Thus young Timmins is ruined through his own speculative greed.

Timmins's precipitate rush to the City on hearing that Hudson had 'joined the line' would have surprised nobody at the time. The paper value of George Hudson's subscriptions lying before Parliament during 1845 was nearly £320,000. His position and influence were impregnable. He was on friendly terms with Prince Albert, and was fêted by the aristocracy, who visited him at his palatial house in Albert Gate, Knightsbridge. He was by now Deputy-Lieutenant of Durham and a magistrate; on 15 August 1845 he was elected Member of Parliament for Sunderland, where he was chairman of the Dock Company.

Thackeray's lively and amusing tales reveal the author's minute observation of the commercial scene, and his detailed knowledge of even minor lines during the period. But, like Dickens's Mudfog essays, they are works of the moment, hastily conceived and published at the height of the Mania, and obviously not regarded by their creator as works of significant or lasting worth.

By 1844 the Government had realized the need to protect the public from self-interested promoters and ill-advised schemes, and in that year set up a Railway Advisory Board under Lord Dalhousie, to scrutinize all new railway Bills. In a single month, May 1845, the Board examined 240 new Bills, with an estimated capital outlay of some £100 million; the total national income for the year ending January 5, 1845, amounted to only £58,760,346. This figure shows the dangerous direction in which the mania was moving.[17]

The mania raged wildly throughout the year as people of all classes rushed to make their fortunes.[18] Internal rivalries led to the sudden abolition of Lord Dalhousie's Board in July, so that there could be

[17] It is indicative of the unreasoning optimism of the period that powerful vested interests attempted to manipulate the Board's decisions, and the very possibility of the Board's finding in favour of a projected line became a subject for speculative share trafficking.

[18] 'Between the months of May and June the increase of speculation was fearful. The papers teemed with advertisements . . . Earls and Marquises struggled with London capitalists and rustic landowners to add attractiveness by the sanction of their names . . . It was a golden moment for the "alley-man", the jackal of the Stock-exchange . . .' (Morier Evans, pp. 5–6).

no control over shady 'bubble' enterprises. The *Bankers' Magazine* warned its readers in September that

> the real object of the concocters of railway schemes . . . [has been] to rob and delude the public by getting their scrip into the market at a premium, and to rob and swindle their subscribers in particular by squandering and embezzling the deposit money.[19]

By this time, George Hudson's personal subscriptions amounted to £818,000.[20]

The inevitable alarm was sounded in October. The Bank's bullion was dangerously depleted, the cost of money suddenly rose, and indiscriminate gambling in railway shares was checked. The paper of established and viable lines held its price, while that of feeble companies fell to a discount. The share market was now free of the enormous masses of worthless paper that had choked it, but the crisis raged on into 1846, in which year George Hudson became for the second time Lord Mayor of York. New schemes continued to be presented, and assent was given to 270 Bills with raising and borrowing powers of £131,713,206. No comparable sum was ever authorized at any other time during the period of private ownership of the railways.[21]

The commercial crisis of 1847 was precipitated only in part by railway speculation. In June 1846, the Corn Laws had been repealed, and the consequent large importation of corn produced an extraordinary excitement in the share market, with an inevitable upward adjustment of the Bank's rate of discount. It rose from 2½ to 3 per cent in 1846, and in the following January to 4 per cent. On 8 April 1847 it reached 5 per cent, with higher rates for long-dated paper. Suddenly, money to back projected lines was not to be had, and shareholders began to sustain heavy losses. The cotton and pig-iron trades were badly affected, and the good harvest of 1847 caused a depression in the corn trade, with numerous bankruptcies. The day was saved when the London merchants petitioned the Government for help. The Government issued a letter of reassurance, in which they

[19] The *Bankers' Magazine*, Sept. 1845.

[20] Morier Evans, op. cit., p. 19 (footnote). Evans here tells us that nearly one quarter of the House of Commons, 157 Members, subscribed for massive sums.

[21] The figures given here are from L. Levi, *History of British Commerce, 1763–1878*, 2nd edn. (1880), p. 303, and from Grote Lewin, p. 115.

advised the Bank to enlarge its discounts on approved security, and to charge an interest rate of 8 per cent.[22]

This had been a particularly severe and distressing crisis, especially for railway shareholders, who had seen a massive decline in the value of railway paper, with £78 million knocked off the value of shares in that year.[23] The magnetic attraction of George Hudson's name now began to lose some of its power to draw, even though public journals, as late as August 1847, still presented the 'Railway King' as ruling supreme.

In *Fraser's Magazine* that month sixteen columns were devoted to Hudson as one of the 'Railway Potentates in Parliament'.[24] The article spoke of his unparalleled influence in the country, his 'singular energy, shrewdness, knowledge, and grasp of mind', and how he was listened to with respect on topics outside his special area, where lesser mortals would have been subject to laughter. While admitting that 'there is no place where success is so worshipped as in the House of Commons', *Fraser's* concludes that even a Mammon-worshipper can be endowed with a special kind of genius. This article gives no hint of the uneasy stirrings that were beginning to undermine Hudson's position. 'Upon railway subjects he is listened to by all parties with respect . . . Here, in fact, he is a positive authority. His decision, *pro* or *con*, on a measure connected with railway management, is almost law.' (p. 221.)

While some literary works appearing at this time made peripheral use of the mania and its resultant crisis,[25] Charles Reade's *Hard Cash* (1863) was deliberately set in 1847, and can thus claim to show more

[22] The letter was issued from Downing Street 25 Oct. 1847, and signed by Lord John Russell, and Sir Charles Wood, Chancellor of the Exchequer. The text of the letter is given in Levi, op. cit., pp. 310–11, and in Morier Evans, op. cit., p. 87.

[23] G. C. Boase, 'George Hudson' in *DNB*, x, pp. 145–7.

[24] (G. H. Francis), *Fraser's Magazine*, Aug. 1847, pp. 215–22.

[25] At the height of the mania, in 1846, Mrs Gore published *Men of Capital*, two distinct tales, one designed to illustrate 'the evil influence of mercenary motives', and the other 'exhibiting the Man of Capital in his nobler phases'. This second tale, 'Old Families and New,' describes the fortunes of Mordaunt, a manufacturer turned railway promoter, who settles in the country, and brings prosperity to all with a new 'railroad'. The neighbours debate his social status, while he falls foul of the decayed 'gentle family' living near by. It is a shrewdly-observed story of social distinctions and public usefulness. In 1849 Catherine Sinclair published *Sir Edward Graham: or, Railway Speculators*, but this is a conventional romantic novel, in which speculation is emphasized only towards the end, in order to furnish a topical dramatic climax. The novel was reissued in 1854, as *The Mysterious Marriage: or, Sir Edward Graham*; the earlier title shows an author hoping to attract her public by hinting at a contemporary theme.

than a passing interest in the events of that year. Richard Hardie, the villainous banker in this novel, had been depicted in the earlier chapters as the saviour of his father's bank during the period of the Company Mania of 1825–7. However, by 1847 he is tempted to play too cleverly in the railway-share market, and to plunge into dangerous speculations.

Reade's epitome of the mania is worth quoting:

> Then railways bubbled. New ones were advertised, fifty a month, and all went to a premium . . . The flame spread, fanned by prospectus and advertisement, two mines of glowing fiction . . . Bishops warned their clergy against avarice, and buttered Hudson an inch thick for shares.' (Ch. 8.)

As always, Reade is accurate in his facts, and in the interpretation that he gives to them. Hardie, for example, advises his customers that the notional capital of authorized lines was £500 million, with a similar sum likely to be required by as yet unregistered projects. If, he said, this money was to come from 'floating capital', then speculators were indulging in the 'Arithmetic of Bedlam' (p. 129). This is a judicious and sound comment, and it is therefore all the more effective to see how Hardie perversely disregards his own commercial instinct, and plunges into time bargains. He stood to gain £30,000 just as the panic began, and prices fell. Charles Reade clearly enjoys the discomfiture of his incredible creation: 'The biter was bit: the fox who had said, ''This is a trap; I'll lightly come and lightly go'', was caught by the light fantastic toe.' (p. 129.)

Reade uses a typical transaction of the mania years to set his protagonist off on a course of quite unbelievable and varied villainies, and while his involvement in the events of the Railway mania are detailed and convincing, Richard Hardie himself remains a memorable, but incredible fantasy.

Badly shaken by the events of 1847, the principal railway companies began to look more closely at details of management, in order to economize, and with the intention of issuing 'general statements' for the reassurance of their battered shareholders. In response to these moves by the companies, shareholders themselves banded together to form committees of scrutiny. Hudson at this time was chairman of four principal railway companies: the Midland; the York and North Midland; the York, Newcastle and Berwick; and the Eastern Counties.

In February 1849, a committee of the Midland Company's shareholders revealed that Hudson, their Chairman, had granted running powers over his York and North Midland line to the Midland's chief rival, the Great North of England Company. No longer afraid to express a decided opinion contrary to Hudson's once unquestionable decisions, the Midland committee asked for Hudson's resignation, which it received in April 1849.[26]

The same month also saw the reports from the shareholders of the other three Hudson companies, all of which revealed grave and irresponsible mismanagement of accounts. The Eastern Counties, originally a struggling agricultural line, had asked Hudson to take control in 1845, and under his aegis it quickly became prosperous. But now, the committee declared that dividends had been paid out of capital, and at a noisy public meeting the resignation of Hudson and the entire Board of Directors was greeted with joyful cheers.[27]

On the following day the committee of the York, Newcastle and Berwick issues their report, which detailed five alleged irregularities committed by Hudson. Share issues had been increased without the shareholders' knowledge, and there had been grave irregularities in accountancy. In particular he was accused of having made a personal profit of £38,000 through deals for the supply of iron rails. Finally, the committee of the York and North Midland revealed that the books of the company had been kept 'in the most slovenly manner', and that since 1845 there had been irregularities in traffic returns amounting to £75,000.[28]

These revelations brought a dramatically abrupt end to George Hudson's dominance over railway finance and development. Called upon to refund over £598,700, his bona fides was all but totally discredited, and between 28 February and 17 May 1849, he resigned from the directorships of all the lines with which he had been connected, or which, indeed, he had created.

[26] 'How Hudson could have reconciled to himself his own position as Chairman of all three Companies when the policy of two of them was directly opposed to the interests of the third and largest Company is a mystery only to be explained by his overweening confidence in his own power to extricate himself from any position in railway affairs, however impossible.' (Grote Lewin, p. 358.)

[27] A verbatim account of this quite astounding meeting is given in Morier Evans, *Facts, Failures and Frauds* (1859), pp. 56–7. On accountancy, note particularly Grote Lewin, p. 353: 'Directors, auditors, and shareholders alike had had no experience of the necessity of depreciation or reserve funds, and widely divergent opinions were held on the subject.'

[28] Morier Evans, op. cit., p. 55. The details of irregularities are fully itemized in Ch. 2 of Evans's work.

Some of the immediate responses to Hudson's downfall were predictable, and indicative of the uneasy relationships obtaining between the classes at that period. Charles Greville, with aristocratic hauteur, wrote that the exposure of Hudson's 'railway delinquency' had caused 'no small satisfaction': 'In the City all seem glad of his fall, and most people rejoice at the degradation of a purse-proud, vulgar upstart, who had nothing to recommend him but his ill-gotten wealth.'[29]

Greville had conveniently forgotten, in writing these cheap sneers, that representatives of his class had been foremost in seeking favours from the 'Railway King': 'Almack's was forsaken when Albert House was full. The ducal crest was seen on the carriage at his door. The choicest aristocracy of England sought his presence . . .'.[30]

The prim City of York erased the fallen magnate's name from the aldermanic roll, and quickly renamed Hudson Street, Railway Street. The less elegant borough of Sunderland, which had benefited enormously from Hudson's exertions, never repudiated him, retaining him as their Member of Parliament until 1859. Ten years later, he was entertained at a public banquet in Sunderland, in recognition of his past services to the town and port.

Two novels of finance that clearly base fictional characters upon the career and personality of George Hudson appeared in the early fifties. One was Robert Bell's *The Ladder of Gold* (1850), and the other *The Gold-Worshippers*, by Emma Robinson, which appeared in the following year.

Bell's Hudson-figure, Richard Rawlings, begins life as an optimistic and honourable young man, but years of drudgery for a mean employer cause him to believe that wealth is the sole means of gaining power over one's fellows. He devotes himself to climbing this 'ladder of gold', but while making his fortune he temporarily loses the affection of his daughters. The story takes place during a period of 'gigantic bubbles', and when the inevitable crash occurs, Rawlings is ruined. Bell contrives a happy ending for his promoter: he creates for himself a more modest fortune as a result of hard work rather than of speculation, and is received once more with affection by his family. There are a number of similarities between the respective careers of Rawlings and Hudson, but the novel offers nothing really original. It is

[29] Quoted in J. W. Dodds, *The Age of Paradox* (1952), p. 386.
[30] The *Bankers' Magazine*, Dec. 1851, p. 749.

6. An amused Mr Punch looks on as Nobility, Clergy, and Army prostrate themselves before the 'Railway King'.

a moral fable of a man corrupted by the power of gold who ultimately forsakes gambling for honest industry, and as such is typical of the tradition of moralistic depiction examined and illustrated at earlier points in the present work.

Similar in many ways is Emma Robinson's *The Gold-Worshippers*. Her protagonist, thinly disguised under the name Humson, also follows closely the general outlines of Hudson's career, and much is made of Humson's lavish entertainment of a sycophantic aristocracy before the inevitable crash.

The image of the promoter as a Golden Calf surrounded by its worshippers occurs often in literature of the period, and is employed by Emma Robinson in her depiction of Humson. 'The idol entered, and all the worshippers were instantly—we cannot exactly say prostrate, except in soul . . . A broad-set, stout, plebeian figure, with a fat, oily visage, a good natured smile, and so corpulent that he waddled as he entered at his brisk pace—and behold the Mammon of our day!'[31]

This is clearly Hudson, and Emma Robinson's description may be compared with the many verbal portraits that appeared in books and journals at the time.[32] Miss Robinson is quick to condemn 'the crawling priests and frequenters of the temple', regarding them as worse than their idol, and this is a timely reminder that gamblers in a time of mania should not be regarded as innocent victims when the crash occurs. However, when Miss Robinson makes Humson 'resort to fraud and deception to keep up the illusion of his power and supremacy', she parts company with her prototype, showing how her fictional promoter is a product of a continuing literary tradition. It would be very difficult to accuse Hudson of any conscious acts of deception, and fraud of any kind was never seriously suggested or maintained, even by Hudson's most inveterate detractors.

In the year that Bell's *The Ladder of Gold* appeared, Thomas Carlyle published his *Latter-Day Pamphlets* (1850), the seventh of which is entitled 'Hudson's Statue'. A sum of £25,000 had been subscribed by admirers of Hudson to commission a public statue of him; however, on his fall from power, the idea was abandoned. Carlyle, who had witnessed the hectic events of the Company Mania of 1825, and then the

[31] E. Robinson, *The Gold-Worshippers* (1851), ii, pp. 27–8.

[32] For instance, in *Fraser's Magazine*, Aug. 1847, p. 222; The *Bankers' Magazine*, Dec. 1851, p. 751; Morier Evans: *Facts, Failures and Frauds* (1859), pp. 70–1.

money-seeking *furore* of the railway era, had few kind words for George Hudson. Offering a variation of the Mammon metaphor, he describes the fallen financier as 'that Incarnation of the English Vishnu',[33] 'the swollen gambler . . . the King of Scrip'.[34] Taking Vishnu largely for granted, Carlyle is more interested in his worshippers. No hasty decision to abandon the erection of a statue will disguise the fact that the subscribers and shareholders concerned contributed as much as Hudson himself to his apotheosis: 'Show me the man you honour; I shall show you by that symptom, better than by any other, what kind of man you yourself are.'

There is much truth in Carlyle's stigmatizing of those who placed Hudson on too high a pedestal; but at one point in 'Hudson's Statue' he makes a remark that brings the reader up with a jolt: 'What Hudson's real worth to mankind in the matter of railways might be I cannot pretend to say . . . From my own private observation and conjecture I should say, Trifling if any worth.' Here the ebullient old wordsmith goes too far, and reveals behind the rhetoric either blind prejudice or genuine ignorance—one prefers to think that it must be the latter. Hudson was fully committed to the development of the railways, and had a minutely detailed working knowledge of the whole system. He was in the true sense of the word an expert, and his concept of amalgamation, taken in conjunction with his creation of the clearing system, has justly earned for him the reputation of having laid the foundations of a unified railway system.

One sees here the dangers of trying to judge the commercial institutions of the era from an examination of fiction and *belles-lettres*.[35] It has led distinguished literary critics to link Hudson with criminal frauds such as John Sadleir in real life and Merdle in fiction, and one hopes that the material presented in this and a previous chapter

[33] Quotations in the text are from the Ashburton Edition of the *Works*, (1893), vol. v.

[34] Scrip, or omnium, as it was sometimes called, consisted of unregistered receipts issued to subscribers, upon which one payment or more had been made. The existence of such a fund of provisional, unregistered securities attracted the unprincipled broker, as it could never be proved whether he actually held the scrip, or was merely buying 'in blank' as a jobbing speculation. The reader will appreciate that, in literary works, the terms 'scrip' and 'omnium' will always have a pejorative connotation, more readily appreciated by the Victorian reader.

[35] Not a few authors were quick to castigate those who eagerly participated in the speculative gamble. See, e.g. Julia Pardoe, *The Poor Relation* (1858), ii, p. 152; W. E. Aytoun, *Norman Sinclair* (1861), ii, p. 33.

has shown how unjust and wrong such identifications are.[36] It is very important in this particular field of study for the critic to be on his guard against becoming part of the literary tradition that he is attempting to analyse.

Charles Greville had said that all in the City had seemed glad at Hudson's fall, but one doubts whether that aristocratic figure had ventured through Temple Bar into the 'horrid City', to ask anyone there for confirmation of his assertion. Had he done so, he might have been surprised at the answers he received.

In the same year that Carlyle's *Latter-Day Pamphlets* appeared, J. A. Francis published his two-volume *History of the English Railway*. In this work, issued only two years after Hudson's fall, one finds a remarkable and dispassionate defence of Hudson from a writer who brought a particularly mordant style of condemnation to bear upon his accounts of the true rogues of commercial life.[37] The *Bankers' Magazine*, in its issue of December, 1851, printed a précis of the relevant passages in Francis's work, together with some important comments of its own, under the title 'Mr. George Hudson and the English Railway System.'[38] Alluding to their précis, the editors state:

> we believe we have presented a dispassionate account of the career of Mr. George Hudson, who is praised for having assisted in developing the railway system as it at present exists[39] . . . although the mad speculation which ensued, and his own conduct subsequently, detracted from the merit of what he had previously effected, *yet we believe he has received a far larger share of censure and vituperation than his faults deserved, while his previous services have been altogether overlooked.* (pp. 746–7. My italics.)

This early vindication of Hudson from the wrong side of Temple Bar appeared in one of the most responsible and highly respected of nineteenth-century journals; it is, though, perhaps inevitable that the noisy rhetoric of Carlyle and the dramatic creations of the novelists

[36] See, for example, Humphry House, *The Dickens World* (1941), Ch. 1, and Grahame and Angela Smith, 'Dickens as a Popular Artist', in the *Dickensian* (Sept. 1971), pp. 131 f.: 'The similarities between Hudson and Merdle are too obvious to be pointed out.' They are not at all obvious, and need to be pointed out, if, indeed, they exist.

[37] See in particular his *Annals, Anecdotes and Legends . . . of Life Assurance* (1853), extracts from which appear in Chapter Four above.

[38] The *Bankers' Magazine*, Dec. 1851, pp. 746–54.

[39] It will be remembered how Carlyle, writing in the same year, dismissed Hudson's contribution as being of 'Trifling if any worth'.

should be remembered and believed, while the works of those who knew the City as it really was should be virtually forgotten.

Francis summarizes the course of Hudson's career soberly and accurately, and incidentally demolishes the idea of Hudson as a shady promoter of 'bubbles' when he asserts: 'his name, moreover, was never connected with a company not meant to be carried out' (p. 749). Although Hudson was one of the 'New Men', and lived in patrician style, he was not a man to forget earlier, humbler friends, and this merit was in part responsible for the fidelity of many former associates towards him in his years of obscurity.[40] 'To his honour he still remembered his friends; he ever enquired kindly after their welfare; he never refused a helping hand to their necessities.' (p. 750.)

Francis makes it clear that Hudson was something more than a company promoter. He was a man of original and inventive genius, a professional manager with a keen interest in his chosen sphere of activity.

> He examined personally every railway department, visited every office, and inquired into the duties of all . . . In railway matters he thus was a director indeed. Not only in the board room, but every letter and every communication bore direction as to some minute detail, which the mass of directors thought beneath them. (pp. 750–1.)

After these necessary correctives to the literary views of Hudson's personal character, Francis embarks upon a careful and detailed consideration of some of the charges made against him at the time of his fall in 1849.[41] In particular he examines the details of Hudson's contract from iron rails made on 11 January 1845, concerning which it had been maintained that he had made a personal profit of £38,000.

[40] In 1863, fourteen years after Hudson's withdrawal from railway business, Dickens caught sight of the former 'Railway King' at Boulogne. In a letter to Forster (*Life*, 1874, edn., iii, pp. 243–4), he described Hudson as 'a shabby man' with a 'desolate manner', whom he could not place. Hudson had just been in conversation with Dickens's friend Manby, who told Dickens that it was Hudson. The novelist asked Charles Manby why he still 'stuck to him', and Manby replied that Hudson 'had so many people in his power, and had held his peace'. Manby responded with friendship to this quality of the fallen capitalist, and clearly knew more about the complex details surrounding his fall than Dickens could possibly have done.

[41] The principal 'charges' have been briefly itemized earlier in this chapter, where they have been presented in the form, and with the interpretations, that appear in the Report of the Special Committee of the York, Newcastle and Berwick Co., 29 February 1849. It will be seen in the text how Francis provides rather different interpretations, which were held by many to be nearer the truth.

Francis shows how Hudson's colleagues had refused to purchase rails at the right market time because they were 'fearful of so great a responsibility', upon which Hudson had personally borne the risk, and had informed his colleagues by letter that he was about to do so. When, at the end of January, the York, Newcastle and Berwick line advertised for 20,000 tons of iron, the price, as Hudson had forecast, had risen cent per cent. Of the quantity of iron required, Hudson provided the line with 7,000 tons at something below the market price. Further details given by Francis show that Hudson had saved the company £7,000, as it was he who induced the contractors, Thompson and Forman, to sell at £12, ten shillings below the market price. The 'profit' of £38,000, in the light of Francis's examination, assumes the form of a fiction of accountancy, and he remarks: 'Well was the question put by a journalist, "Were the company losers? No. Was the price above the market price? No. Was it bought in the name, or on account, or at the will of the company? No. If iron fell, who would be the loser? Mr. Hudson!"' (p. 752).

Space forbids a full discussion here of Francis's article, but the reader is referred to the convenient epitome in the *Bankers' Magazine*, and to the two volumes of Francis's *History of the English Railway* for further details of these more favourable interpretations of the momentous transactions.[42] The article concludes with the statement that Hudson was guilty of a lack of discretion rather than a lack of principle. Undoubtedly a man with an excellent conceit of himself, he was careless in his accounting, and certainly culpably negligent.

A similar apologetic tone was adopted eight years later by Morier Evans in his *Facts, Failures and Frauds* (1859). Although Hudson, 'much to his credit', had repaid large sums and made arrangements for others, he was attacked with 'such a spirit of hostility' that 'every conceivable method was attempted to crush the last vestige of his popularity' (p. 65). Despite this, and the loss of all his possessions and estates, sold to recoup funds misapplied by his negligence, he 'sustained himself with an apparent amount of fortitude which was truly astonishing, and, except to those personally acquainted with him, almost passing belief.' (pp. 66–7.)

[42] In this connection the reader is referred further to the letter of defence sent by Hudson to the Committee of the York, Newcastle and Berwick Co. It is reproduced in full in Grote Lewin, pp. 360 f.

No company with which Hudson was associated ever failed, ruining shareholders and investors; when called upon to make restitution of sums of almost unbelievable magnitude in terms of modern values, he did so, at total sacrifice of his property and fortune. To those who investigate the details of Victorian finance, there is much to respect and admire in this fallen magnate.

For several years after his fall, Hudson attempted to engage in financial operations on the Continent, but these appear to have been unsuccessful. In 1868 a group of friends raised a subscription of £4,800, and purchased an annuity for him. In the following year he was entertained at a banquet in Sunderland, which he had represented in Parliament from 1845 to 1859. He died, aged 71, at 37 Churton Street, Belgrave Road, London, on 14 December 1871.

On the Saturday after his death *The Times* printed a long obituary of the former Railway King. After nearly thirty years, the dubious moral outrage of the late forties could be viewed in a more critical light. The paper castigates the people who had 'cringed to him and flattered him' during the height of his power, and who had then 'avenged itself by excessive and savage reprobation'. The anger of these people was increased when it was discovered that 'he had really nothing left'. He was a man 'who united largeness of view with wonderful speculative courage'.[43]

On 21 December 1871, George Hudson was buried in Scrayingham churchyard, Yorkshire. It is significant that the City of York had so far relented its earlier pettiness as to send the Lord Mayor to the funeral, and to cause the great bell of York Minster to be tolled. The cortège left York station at 9.30, attended by a great number of people, and shops along the route were closed; shops were also closed at Whitby and Malton. Present at the funeral were many old railway friends and acquaintances of former years.[44] It is doubtful whether Mr Merdle's body was greeted with such respect on its journey to the grave.

[43] *The Times*, Saturday, 16 Dec. 1871, p. 9. Gladstone had called him 'a very bold, and not at all unwise, projector' (Grote Lewin, p. 365). Of modern judgements, that of J. W. Dodds is among the most well-informed and judicious: 'Many responsible journals pointed out that Hudson was as much the victim of his former sycophants as of his own naïve disregard for financial proprieties . . . At this distance there is no little pathos in the career of the Yorkshire linen draper who was little worse or no better than the railway gamblers for whose greed he had become the symbol.' (*The Age of Paradox* (1953), p. 386.)

[44] *The Times*, Friday, 22 Dec. 1871, p. 3.

7. The Spider's Web.

The examination of the life and career of George Hudson given here confirms the admonition made earlier about the relation of fact and fiction. The success of a fictional character depends ultimately upon the author's creative skills, and it is upon these alone that a character is usually judged. However, the personal prejudices of an author, and the influence of a particular literary tradition, can often obscure the real truth when a character is claimed, overtly or by implication, to be a reflection of real life. The student of nineteenth-century fiction needs to examine carefully and impartially the real-life prototypes of such characters as Emma Robinson's 'Humson', Bell's Richard Rawlings, Dickens's Merdle, or Disraeli's Sidonia, if he is not to allow his own estimate of nineteenth-century business figures to be over-influenced, or even formed, by the creations of the novelists.

9

Transmuting the Commonplace: a Problem of Style

The novelist cannot avoid making a 'milieu', an 'entourage' for his characters, and it often occurs that the characters themselves are made secondary to and illustrative of particular social conditions. . . . The objects which interest the mind are determined accordingly . . .[1]

These words of the nineteenth-century literary critic D. G. Thompson alert the reader to a truth that can easily be overlooked. In real life, people need not be intrinsically bound to their setting, but in fiction character and milieu are part of a creative unity, carefully manipulated by the novelist for his own inner purposes. In 'social' novels, as Thompson suggests, the characters may become subordinated to their setting, but whatever the genre of writing, there must always be a predetermined relationship between the figures who move the novel's plot and the entourage in which they find themselves, a relationship that is developed by the controlling intelligence of the novelist.

The realism or otherwise of the surroundings in which the character is placed depends ultimately upon the author's didactic intentions: his desire to illustrate a theme or idea that inspires him also controls his selectivity. Details of setting can, perhaps unconsciously, be 'edited out' if they happen to conflict with the underlying thesis. At the same time, a commonplace reality can be transmuted into 'something rich and strange' to make it accord with an overmastering theme. This is certainly true of many novels of finance, and in this final chapter we shall examine some of the ways in which nineteenth-century authors manipulated the everyday realities of commerce to create the essential 'milieu' or 'entourage' for their characters. The scope of our exploration will be widened somewhat at this point in the book to include some fictional reflections of members of the merchant and manufacturing classes.

[1] D. G. Thompson, *The Philosophy of Fiction in Literature* (New York, 1890), pp. 131–6.

A classic illustration of a commonplace milieu significantly changed for literary purposes is found in Disraeli's *Coningsby*, where, in Book IV, chapter 3, the youthful hero finds himself in the industrial north of England. This area is described in terms designed to present images of vitality and activity, because Disraeli's selective eye is seeking principles of vigour and creativity in the new industrial society that he can ally with the questing, venturesome spirit of the 'new generation' of young aristocracy typified in the novel by Coningsby himself. This desire for a union of abstract principles controls Disraeli's descriptions, and dictates the quality of his style.

The rather unprepossessing area around Manchester and Bolton is thus changed into 'plains where iron and coal supersede turf and corn'; the antithesis is there to symbolize the polarities of land and capital, which Coningsby is to unite by marrying Edith, daughter of the mill-owner Millbank. From this moment in the novel Disraeli, either consciously or otherwise, is concerned that there should be nothing ignoble in the dowry that Edith will bring to his hero. Thus the factories and the 'hands' are described in terms connoting elegance and contentment. We are told of 'illumined factories, with more windows than Italian palaces, and smoking chimneys taller than Egyptian obelisks'. 'He entered chambers vaster than are told of in Arabian fable, and peopled with habitants more wondrous than Afrite or Peri.'

The images of palace and obelisk, the magical Arabian Nights ambience, are part of a desire to soften the harsher realities of the weaving-shed, which is to become part of the noble Coningsby's inheritance. The same selectivity informs Disraeli's depiction of the working people: 'Does not the spindle sing like a merry girl at her work, and the steam-engine roar in jolly chorus, like a strong artisan handling his lusty tools and gaining a fair day's wages for a fair day's toil?' A utopian scene, notably omitting to mention the thunderous clacking of the shuttle-boxes, that could bring on deafness after a year in the sheds. The mill-girls, too, belong to another world, 'in their coral necklaces, working like Penelope in the daytime . . .'.[2]

[2] When Disraeli the politician and social prophet replaces the visionary for a moment, the romantic imagery yields to sober and astute prose: 'this wealth was rapidly developing classes whose power was imperfectly recognized in the constitutional scheme, and whose duties in the social system seemed altogether omitted'. In an overtly political novel that needed to be said; but the analytic tone is soon dropped as the author goes on to elevate the commonplace for thematic reasons.

When we come to examine Coningsby's relationship to this idealized setting, we find a corresponding air of romanticized unreality. The palaces and obelisks are there because Coningsby is no humdrum railway passenger journeying from London to Manchester. He is in the North for an almost preternatural reason: 'Because a being, whose name even was unknown to him, had met him in a hedge ale-house during a thunderstorm and told him that the age of ruins was past.' He is Sidonia's disciple, sent forth into the unknown to learn what has replaced the age of ruins, and like all such venturers sent forth at the behest of sages, he must gaze in wonder at marvels; the harsh and not very beautiful realities of mill life are of necessity here excluded from Coningsby's sustaining entourage.

It could be argued that the sudden appearance in this chapter of the down-to-earth business man G. O. A. Head shatters the fairy-tale quality of what has just preceded it. However, when this vignette is examined, we see how Head is presented as a humanizing of the vital force hitherto confined to engines and machinery. Sidonia had been revealed to Coningsby almost in the manner of an epiphany, and in the same way G. O. A. Head, restless and rather *distrait*, catechizes the young seeker after truth when they two are the sole occupants of a hotel coffee room at night, in an area where Coningsby is a total stranger. Head does not appear again in the novel, but, like Lord Minchampstead in Kingsley's *Yeast*, he functions as an illustrative assertion of the vigorous activity associated with the creators of the Industrial Revolution.

We do not follow the fortunes of G. O. A. Head, as to do so would plunge us into the kind of mundane reality that would damage Disraeli's purposes in this chapter. Instead, we accompany Coningsby into the domain of another mill-owner, Mr Millbank. The reader has been made aware of this character earlier in the novel as a man of 'a very democratic bent', who had 'sent his son to Eton, though he disapproved of the system of education pursued there, to show that he had as much right to do so as any duke in the land.' He is able to do so because he is 'one of the wealthiest manufacturers in Lancashire'. We learn, too, that he has a 'prejudice against every sentiment or institution of an aristocratic character'.[3]

With such qualifications, we could be forgiven for expecting Millbank to be cast as the embittered rich parvenu rejected by the

[3] Quotations are from *Coningsby*, i, Ch. 9.

landed interest, like Colonel Bramleigh or Mr Mordaunt. Admittedly, he is the sworn foe of Coningsby's kinsman Lord Monmouth, but it is significant that Disraeli keeps both men in their own worlds. Millbank remains content to be proprietor of his own northern domains. His creator, true to his thematic intention, clothes his mill-owner with an aristocratic entourage that belies his apparently democratic character. His very mill is 'not without a degree of dignity', and his house is sufficiently impressive, 'built in the villa style, with a variety of gardens and conservatories'.[4] There is no hint of the parvenu, no 'flashiness', as Coningsby and Millbank enter 'a capacious and classic hall'. Here Coningsby is introduced to Edith, a girl of 'peculiar beauty', and her father tells the young man proudly that she has a Saxon name, 'for she is the daughter of a Saxon'.

From this time forward it is easy to forget that the Anglo-Saxon father and his constantly bowing and blushing daughter are in reality on a social par with G. O. A. Head, or the rather gauche John Thornton in Mrs Gaskell's *North and South*. And this is how it must be, for in the end Coningsby and Edith must forge the alliance of aristocrat and bourgeoise, and reside in noble splendour in Millbank's hall, 'in a manner becoming its occupants'. The realization of that alliance has dictated the style and imagery of the chapter, transmuting the unlovely environs of Manchester and Bolton into a midsummer night's dream peopled with happy mechanicals, and has clothed the democratic 'new man' with all the essential ethos of aristocracy.[5]

The reader will realize how fruitless it would be to regard Millbank as a faithful reflection of a genuine northern business man, and the same can be said of many commercial figures in novels where the author's fiction is designed in part to illustrate a strongly-held conviction. Dickens, as we have seen, shows careful and minute knowledge of many commercial transactions, but there are times when he allows a thematic purpose to do violence to the realities underlying his business figures. This is particularly true of his depictions of the

[4] Disraeli's description contrasts tellingly with the satiric picture of 'Zero Lodge', a pseudo-aristocratic dwelling in M. J. Higgins's '*Jacob Omnium. The Merchant Prince*' (1845): 'a sort of parody on the luxurious accessories of a well-appointed country seat. There was a fish-pond, a conservatory, a summer-house, a pheasantry, a dog-kennel', etc.

[5] A much more down-to-earth view of alliances between aristocrat and bourgeois will be found in *The Bramleighs of Bishop's Folly* (1868), where it is said of the old and impecunious Lord Culduff: 'Such a man has only to go down into the Black Country or among the mills, to have his choice of some of the best-looking girls in England, with a quarter of a million of money' (Ch. 18.).

mercantile classes, where the commonplace realities of their pursuits are not permitted to hold the stage for very long.

Dickens creates four fairly well-developed companies in his works, Cheeryble Brothers, Anthony Chuzzlewit & Son, Scrooge & Marley, and Dombey & Son. In all these firms there is an initial suggestion of reality, followed by an almost unconscious shaping into a mould determined by Dickens's preoccupation with theme or character. It is educative to examine these fictional companies in detail, in order to see what this great creative artist made of the realities of business life in the mercantile sector.

Cheeryble Brothers, Dickens tells us in chapter 35 of *Nicholas Nickleby*, are German-merchants, and at certain points in the novel we are reminded of this: we learn in chapter 43, for instance, that Frank Cheeryble had for four years superintended the business in Germany. The Cheerybles, then, were importers and exporters in the German trade, and at the time the novel was written would have been mainly concerned with the import of fine wool from Saxony, and the export to Germany of finished textiles, cotton twist, and yarn. Germany was England's best customer for worsted, used for light fabrics, providing the Bradford manufacturers with profitable, steady, if unexpansive markets.[6] It is interesting to note that Frank has for six months been establishing an agency in the north of England (Ch. 43). The general concept of this early Dickensian firm is thus accurately presented, and, in structure at least, affords a telling contrast to Ralph Nickleby's speculative 'bubble' enterprise, the United Muffin Company, in the same novel.

The direction that this firm takes in the novel is determined partly by the real-life brothers upon whom the Cheerybles were based, and partly by the need to balance the evil Ralph Nickleby with an embodied power of good. Beyond both Ralph and the Cheerybles is the motivating force of this early novel, Dickens's contribution to the age-old theme of the unending war between good and evil. In the later months of 1838, while *Nicholas Nickleby* was in progress, Dickens took a holiday in which he travelled through some of the industrial districts of England. In the Birmingham area he travelled through 'miles of cinder paths and blazing furnaces and steam-engines—a mass of dirt

[6] J. H. Clapham, *An Economic History of Modern Britain* (1926), 2nd edn. (1930), i, pp. 243 and 481. From about 1836 Bradford began to spin alpaca wool from South America, and this proved highly profitable. See, e.g. Clapham, ii, p. 224. Saxon wool was finally replaced by that of Australia, New Zealand, and South Africa.

and misery',[7] and in Manchester he had been horrified by the wretched conditions in the cotton mills. 'I went to Manchester and saw the *worst* cotton mill and then I saw the *best* . . . There was no great difference between them . . . what I have seen has astonished and disgusted me beyond all measure.'[8] Dickens, whose eye always sought out the lowly, saw no lusty, singing workmen, palaces or obelisks, but only 'unfortunate creatures' whose plight he determined to reveal at some future date in his writing.

On the same visit to Manchester, however, he met the philanthropic manufacturers Daniel and William Grant, who carried on business at Cheeryble House in Canon Street. Enormously impressed by their benevolence, he sings their praises in the preface to *Nicholas Nickleby*, and transmutes them into the Cheeryble Brothers. His disgust at what he saw in Manchester may have determined him to distance the brothers from all direct connection with that town, by depicting them as German-merchants,[9] and it is possible that a similar recollection caused him to depict the villainous Anthony and Jonas Chuzzlewit as Manchester warehousemen.

It is significant, though, that the brief business detail in the novel centres on the young nephew Frank, whose partner Nicholas will ultimately become. The brothers themselves are used entirely as a moral counter to the usurious and parasitic Ralph. Their benevolence, as Bagehot and Gissing both found, becomes tiresome, and one is saturated by the relentless string of adjectives used to describe them in chapter 35; sturdy, dimpled, honest, merry, happy, pleasant, comical, jolly, engaging. The exaggerated exercise of these qualities, so necessary to point the contrast with the sinister Ralph, stretches the credibility of brother Charles and brother Ned as business men to the limit. The terms of Nicholas's engagement are sufficiently astounding.

[7] Una Pope-Hennessey, *Charles Dickens* (1945, Reprint Society edn. 1947, p. 106). The deep impression made on him in this area took literary form later in some powerfully atmospheric description in *The Old Curiosity Shop*.

[8] Letter to Edward Fitzgerald, quoted in E. Hodder, *Life of the 7th Earl of Shaftesbury*, i, p. 227.

[9] Philanthropy in Manchester was not limited to the Grants. Mention may be made of the radical merchants Sir Thomas Potter and George Wood, whose work is fully described in Fox Bourne's *English Merchants*, Ch. 19: 'All the world knows the portrait of the Brothers Grant in *Nicholas Nickleby*, there termed the Brothers Cheeryble; and scores of others, less modest and reticent, have made for themselves as great a fame, without the aid of fiction.' (Ch. 19, p. 387.)

'He has done it!' said Tim . . . 'His capital B's and D's are exactly like mine; he dots all his small i's and crosses every t as he writes it. There a'nt such a young man as this in all London . . . not one. Don't tell me! The City can't produce his equal. I challenge the City to do it!' (Ch. 37.)

This nonsense, coupled with the almost imbecilic emphasis upon the neatness of the firm's books, does not echo the realities of nineteenth-century business life. Neither can we be impressed with the repetitive, non-progressive nature of Tim Linkinwater's daily round as depicted in chapter 37, none of it being typical of counting-house practice at the time. Tim would not have survived long in the very active ambience of a London German-merchant's establishment.

Functionally the Cheerybles are the means of rescuing Nicholas from penury at a point when the novel is more than half finished. Their role as merchants is of little importance in comparison with their qualities of fairy godfathers, the righters of wrongs, symbols of light and integrity in the dark and twisted world of Ralph Nickleby and Arthur Gride, a world that ultimately they are able to destroy. Thus they have no qualities of the 'new men', and would be at home with Mr Wardle at Dingley Dell. Dickens leaves his contemporary readers to think of contrasts to the Cheerybles from their own observation of the industrial age, just as later he will make such contrasts overt, as he does with Montague Tigg and Mark Tapley. The underlying themes of *Nicholas Nickleby* and *Coningsby* are very different, but the fact that each has a formative impetus is sufficient to influence character-function, and setting, in ways that become radically different from commonplace realities.

Anthony Chuzzlewit & Son are Manchester warehousemen, distributive merchants and export factors for the Lancashire cotton goods and Yorkshire linens that emanated from the bustling manufacturing establishments of Manchester, which had so impressed young Coningsby. They would thus be participants in a constantly thriving trade. Firms like Chuzzlewit's would have received regular consignments of goods from such firms as Thomas Potter's of Canon Street, which occupied premises six storeys high, used steam hoists to load its vans, and had fifty men constantly engaged in packing and unpacking bales.[10] In 1843, the year in which *Martin Chuzzlewit* was published, there were some 360 mercantile houses in Manchester, most of which provided thriving business to houses and agents in London.

[10] Detail from J. G. Kohl, *England and Wales* (1844), pp. 126–7.

It is quite impossible to equate Anthony Chuzzlewit & Son with this particular business ethos. Their premises occupy 'a dim, dirty, smoky, tumble-down, rotten old house', in the gloomy chambers of which are to be found 'linen rollers, and fragments of old patterns, and odds and ends of spoiled goods, strewn upon the ground'. (Ch. 11.) This detail suggests that the Chuzzlewits, like the Fezziwigs in *A Christmas Carol*, live on their premises, an old-fashioned practice by 1840. There is little of the 'new man' about old Anthony: he actually belongs to the same old order as the Cheerybles. But whereas Nicholas's patrons and Scrooge's old master symbolize joyful life and honest commercial enterprise, Anthony Chuzzlewit & Son is redolent of commercial death and decay.

The mouldy tatters of Manchester-goods inspire no confidence in the father and son as merchants or factors, and, as so often in Victorian novels, we are never shown these two exercising the mysteries of their trade. And yet there is a kind of hideous vitality about the Chuzzlewit ménage, and Dickens tell us in chapter 11 that 'Business, as may be readily supposed, was the main thing in this establishment.'

It is here that we begin to perceive how Dickens's proclamation in this novel of the evils of Mammon—the word is uttered on one occasion with shuddering hypocrisy by Mr Pecksniff—has determined the entourage of these characters. Their professed business must be shown in decay as a symbol of their moral degradation, their determination not to indulge in 'honest toil'. In its place, Dickens substitutes another kind of 'business', that of usurious and grasping money-getting, which places the father and son in the world of Arthur Gride and Ralph Nickleby.

Like Ebenezer Scrooge, the Chuzzlewits are dedicated to the 'pursuit of wealth'.[11] This will be effected not through mercantile dealings but through swindling—'Do other men, for they would do you' is Jonas's business credo—particularly through associating with the swindler Tigg in his bogus insurance enterprise. The suggestion of usury becomes an open fact when, towards the end of the novel, Lewsome makes his confession: 'We met to drink and game; not for large sums, but for sums that were large to us. He [Jonas] generally won. Whether or no, he lent money at interest to those who lost . . . he came to be the master of us.' (Ch. 48.)

[11] Ch. 11. This same expression, 'the pursuit of wealth', is used to describe Scrooge's ambition in *A Christmas Carol* (p. 79). Page refs. in the text are to *Christmas Books* (Penguin English Library), i, 1971.

This has been the controlling idea in Dickens's mind from the beginning, an equation of money-getting with usury, and the generally detested trade of moneylender. Thus it is that the Chuzzlewits are provided with a milieu from which they cannot escape, so reminiscent of the wretched, decaying premises of Arthur Gride, whose spiritual brethren they are: this literary purpose makes it well-nigh impossible for Dickens to create a realistic portrait of men engaged in the busy but inevitably prosaic Manchester-trade.

Dickens does not state explicitly the business pursuits of Scrooge & Marley in *A Christmas Carol* (1843), but as we are told that the firm possessed a counting-house (p. 47), and that old Marley's name remained after his death painted 'above the warehouse door' (p. 46), it is clear that they were merchants of some sort. Further information is provided in the opening paragraphs of Stave Two, where Dickens imagines an example of the kind of paper with which Scrooge would be familiar: 'three days after sight of this First of Exchange pay to Mr. Ebenezer Scrooge or his order . . .' (p. 66). Apart from one small slip this is the wording of a foreign bill of exchange; this one would have been made out by Scrooge himself and sent to a debtor abroad. The slip mentioned involves the time limit. No foreign bill could have been payable three days after sight, and I suggest that Dickens meant to write 'three months', which was a frequently allowed acceptance period.[12] From this detail we can deduce that Dickens envisaged Scrooge as an exporting merchant, perhaps in general commodities.

Scrooge's sins are those of omission, and he will be readily redeemed under the influence of the Christmas Spirits, and it is therefore apt that Dickens should make him a merchant, as they were untainted by the usuriousness attached in the literary mind to usurers, jobbers, and the like. We shall see, though, that the hint of the moneylender will attach itself to Scrooge as it did to the Chuzzlewits, even though Dickens takes pains to depict Scrooge as being commercially impeccable.

Scrooge is certainly a cold-hearted miser, 'a squeezing, wrenching, grasping, scraping, clutching, covetous old sinner!' (p. 46.) These are unpleasant characteristics, but they do not necessarily connote

[12] Foreign bills of exchange were prepared in sets of three, to avoid loss in transit. These copies were known as the First, Second, and Third of Exchange. In later practice, to avoid the possibility of duplicate payment, the words 'Second and Third not paid' would have been added to Scrooge's bill. The expressions 'First of Exchange', etc., were never found in an inland bill.

dishonesty: 'Scrooge's name was good upon 'Change for anything he chose to put his hand to', and this vouches for his business integrity.[13]

The reader, however, is not seeking for these commercial details, nor does Dickens insist upon them. Instead, he emphasizes Scrooge's cupidity and avarice, so that, after his conversion, his new-found bounteousness will strike us with all the more force and delight. It is not long before the shades of the moneylender's den begin to gather stylistically, and the honest merchant's counting-house is transmuted into something significantly different.

The feeling of usury, of dealing in money rather than in commodities, pervades the story from the beginning. The items that constitute Marley's chain are primarily suggestive of miserly hoarding: cash-boxes, deeds, 'heavy purses wrought in steel' (p. 57). The spirit of his old partner tells Scrooge: 'in life my spirit never roved beyond the narrow limits of our money-changing hole' (p. 61). The expression 'money-changing' is significant in that it directly connotes the moneylender, and could also conjure up in the Victorian mind some of the dubious activities of the money market. The expression shifts Scrooge and his firm very effectively away from their true identity as merchants.

As Dickens's creative imagination dictates his style and choice of words, the images of the usurer continue to be attached to the 'covetous old sinner'. The last of the spirits conducts him to a house in which a woman is anxiously awaiting her husband. She seems to be in straitened circumstances, as she has been 'hoarding' his dinner by the fire. When her husband arrives with news of Scrooge's death, she asks him: 'To whom will our debt be transferred?' He does not know, but asserts that 'it would be bad fortune indeed to find so merciless a creditor in his successor'.[14]

Earlier, this man had attempted to secure 'a week's delay' for payment of his debt. Dickens does not mention this man's profession, and it may be that we are to understand him to be a retailer who has failed to meet the due date of a bill. He is unlikely to be a fellow-merchant, as he would almost certainly have followed accepted practice

[13] p. 45. ''Change' in this context is, of course, not the Stock Exchange, but the Royal Exchange, where merchants and bankers frequently met to transact business. Scrooge probably frequented the second Exchange building, which was burned down on 10 Jan. 1838. The present familiar building, between the Bank and Cornhill, was opened by Queen Victoria on 28 Oct. 1844.

[14] Ed. cit., p. 120.

by seeking discount on his bill at his bank or discount house before the bill fell due.

The domestic setting of this little incident, the wife's expression 'our debt', the week's delay sought by the husband, and the expression 'so merciless a creditor' all connote the private loan at high interest of the moneylender. With Scrooge & Marley, as with Anthony Chuzzlewit & Son, we see how Dickens starts with a merchant house, and almost unconsciously transforms it into a Temple of Mammon, with an unmistakable flavour of the usurer and extortioner. The demands of his splendid moral fable require him to do violence to Scrooge's ostensible profession.

When Dickens created the firm of Dombey & Son, 'wholesale, retail and for exportation', he presented himself with certain difficulties in establishing a satisfying 'entourage' for his protagonist. In the person of Mr Dombey Dickens is concerned not with a bona fide portrait of a merchant, but with the soul-sickness of a man made arrogant by wealth. Despite the title of the book, which is ostensibly the name of a firm, *Dombey and Son* is essentially a domestic novel, examining and developing the relationship of a man and his daughter: 'not so much a "business" novel as a book about a proud rich man who refuses, like Scrooge, to respond to human requirements of love and kindness. . .'.[15]

It is essential to the working out of the theme that Dombey should be rich, but at the same time it is equally essential to realize that he is never in any sense a Mammonite: throughout the novel there is no hint of any greed for gain on his part. Like Scrooge, he is in essence an upright and honourable business man; unlike him, he is no miser.

With these qualities of character in mind, Dickens, initially at least, chooses Dombey's profession with considerable care. From the fact that Dombey has an agent in Barbados, and from the visits paid by Sol Gills to that island and to Jamaica and Demerara, it becomes clear that Dombey is a West India merchant. Dickens thus places him in the most distinguished rank of merchants, enjoying a power and prestige far greater than that of any other class of trader.[16] The trade

[15] R. H. Dabney, *Love and Property in the Novels of Dickens* (1967), p. 55. Dabney argues that Dickens's preoccupation with Dombey's marriage to Edith in the second half of the novel turned its course into a shape not originally planned.

[16] The West Indies played a prominent role in England's external trade. In 1830, for example, £2,800,000 worth of goods were exported to the islands (Clapham, i, p. 250). Speculation in West India produce received much attention from produce-brokers, who

was old and stable, and not in competition with that of the mother country; Dombey, who could have been shipowner as well as merchant, would have shared in the very profitable export business to the West Indies by shipping out indigenous goods, particularly Manchester wares, offloading them in the Caribbean, and filling his holds with raw sugar.

He would, too, have belonged to the powerful 'West India Interest', a protective grouping of merchants and planters that had arisen during the eighteenth century. The Interest was supremely successful in keeping up the demand for West Indian sugar by exerting political influence against rival markets, and often swaying the policies of the Government.[17]

It is very possible that Dickens had these qualities of the West India trade in mind when creating an ambience for Dombey, because the stability and prestige of this branch of commerce could conceivably have generated the kind of complacency over business matters, and the incautious sense of security, that are to lead to Dombey's commercial ruin.

One can detect a growing lack of interest in Dombey as a businessman from as early as July 1846, when the first four chapters of the novel were still in manuscript. In a letter to Forster written on 25 July, Dickens says: 'So I mean to carry on . . . through the decay and downfall of the house, and the bankruptcy of Dombey, and all the rest of it . . .'.[18] This intention was never carried out: the illustration of a gradual decline suggested by the word 'decay' does not materialize, and the careless phrase 'all the rest of it' indicates that his mind was elsewhere than in the counting-house while he was considering what to make of his proud merchant. And so it is that Dombey, having been placed by his creator in the top rank of the

used speculators' money to corner the market in a commodity, realizing large profits for themselves and their investors by gradual release of the product on to the open market. The process is described in Samuel Warren's *Passages from the Diary of a Late Physician* (1854 edn., pp. 231–2).

[17] Details of the Interest given here are derived from L. M. Penson, 'The London West India Interest in the Eighteenth Century', *English Historical Review*, July 1921, pp. 373–92. The only commodity mentioned in *Dombey and Son* is 'hides': 'Dombey and Son had often dealt in hides, but never in hearts' (Ch. 1). I think that Dickens chose 'hides' simply to alliterate with 'hearts'. The leather trade, medieval in practice and untouched by the Industrial Revolution, could offer little to a merchant of Dombey's calibre (Clapham, i, pp. 170, 234–5).

[18] Forster, ii, p. 312.

merchant classes, proceeds to idle his time away for over three years of the novel's time-scale, neglecting the business in order to concentrate on his private life—at least, that is what *appears* to be happening. In fact, Dombey is simply conforming to what many critics see as an abrupt change of plan by Dickens, in which the author's preoccupation with the themes of Dombey and Florence, and then Dombey and Edith, move the novel in a direction not originally intended, so that Dombey, yoked to his original 'entourage', seems wilfully to neglect his business when in fact his creator has left him no option.[19]

It is not, therefore, surprising that in this novel the depiction of a firm transacting business is particularly vague and unimpressive. Dickens has no real interest in the dealings of the firm, and what details do emerge are mere window-dressing. In chapter 13, for instance, Carker is depicted 'with a bundle of papers in his hand', and in chapter 23 we find him again 'pausing to look over a bundle of papers in his hand'. In chapter 46 he explores 'the mysteries of books and papers' in the strong-room. After the crash, we are told how the accountants unveiled 'the great mysteries, the Books' (Ch. 58). Mysteries to us, and, one suspects, to Dickens also!

Perhaps the ultimate embarrassment for anyone trying to make Dombey accord realistically with his high mercantile calling is to be found in the following piece of dialogue between Dombey and Carker:

> 'What business intelligence is there?' inquired the latter gentleman after a silence, during which Mr. Carker had produced some memoranda and other papers.
> 'There is very little,' returned Carker. 'Upon the whole we have not had our usual good fortune of late, but that is of little moment to you . . .' (Ch. 26.)

What merchant in real life would agree with his manager that a fall in fortune was 'of little moment' to him! Dombey, of course, is constrained to agree because of what Dickens is doing to him, but his complacency in this episode makes nonsense of Joe Bagstock's description of him as 'the Colossus of Commerce'.

Dickens is concerned to show how Dombey's native wit is clouded with his own conceit, so that this absence of astuteness and acumen may be in part deliberate. The faithful Mr Morfin is forced to admit that his master positively refuses to believe that the House 'is, or can

[19] See in particular *Dickens at Work* (1957), in which Dr Butt and Dr Tillotson comment on the almost total lack of 'dealings' with Dombey's firm.

be, in any position but the position in which he has already represented it to himself.' (Ch. 53.) Even without Carker, such a principal would surely have ensured the very swift downfall of Dombey & Son.

Illustrations of this kind could be multiplied from the considerable number of Victorian novels that concern themselves with prevailing commercial and mercantile institutions.[20] In most cases, of course, it would be churlish to regard the various inaccuracies as blemishes, as few novelists would wish to be regarded as writers of commercial textbooks. Nevertheless, it is important for the modern reader to realize that the creative minds of the novelists make selectivity and transmutation of reality virtually inevitable.

Of all nineteenth-century novelists who turned their attention to the City and its doings, Mrs Gore was the most faithful to reality, and the best able to solve the problems of presentation caused by the demands of structure, setting, and style. Virtually forgotten today, her works were enormously popular in her own time: such disparate critics as George Eliot and George IV thought highly of her talents. Her success in depicting the financial scene was due partly to her innate ability, and partly to the nature of her formative years as a writer, which developed in her a broad view of society and its doings, and a certain wariness about the antipathetic literary tradition in the matter of the City inherited from the eighteenth century.

Born in 1799, she married in 1823 Captain Charles Arthur Gore, an officer in the 1st Life Guards, and from the time of her marriage pursued a career in writing and publishing. She was an excellent woman of business, though, as we have seen, singularly unfortunate in her choice of banker. Her literary output was phenomenal, amounting to some seventy different works extending to nearly two hundred volumes, in which, according to the *Gentleman's Magazine*, 'there is scarcely to be found one dull page'.[21]

[20] Of particular interest in this respect are the curious business methods of Mr Thornton in Mrs Gaskell's *North and South* (1855). With a large stock of raw cotton lying idle in his stores, he wastes much time trying to 'detect the secret of the great rise in the price of cotton'. Such a rise would have had the effect of increasing the value of his store, which he could have sold at a great profit. However, it suits Mrs Gaskell's literary purpose that Thornton should fail. In his search to find out why the cotton price had risen, he journeys to Le Havre. The chief cotton market of the world was Liverpool, at the other end of the Liverpool to Manchester Railway. His journey to Le Havre was thus clearly unnecessary, as England imported through Liverpool four times as much cotton as did France.

[21] *Gentleman's Magazine*, Mar. 1861, p. 346 (Obituary notice).

Mrs Gore's forte was the writing of novels of 'fashionable life', a genre that was popular in the early decades of the century. But, although Thackeray parodied her style in *Lords and Liveries*,[22] thus causing some critics to regard her as one of the 'silver fork' school, she was never so regarded in her own time.

> She satirises, as well as depicts, the gay world. She shows it, and something more—she shows it up. She does not require us, as the true fashionable novelist does, to fall down and worship her image; nay, she bids us rap our knuckles on its brow, and mark the echo of sounding brass . . .'.[23]

Mrs Gore was a natural satirist with a delightful and trenchant wit, who exposed the follies of her upper-class characters, and was not above including her own chosen mode of writing in the catalogue of trivia associated with them: 'those dreadful, flimsy, unwholesome tissues of false sentiment and flippancy, called fashionable novels, were composed for the delight of the bankers' wives'.[24]

Mrs Gore's literary reputation was firmly established by the appearance, in 1841, of *Cecil, or the Adventures of a Coxcomb*, which created a genuine sensation in the literary world and with the reading public when it appeared. *Cecil* is the story of a young man who is himself a satirist of the *beau monde*, and at the same time an abject slave of its fashions and frivolities. It is an enormously clever and entertaining work, exhibiting Mrs Gore's ability to stand back from her subject and observe it with judicious detachment. It is this quality that she brought to bear upon her novels of finance when she turned her attention to the theme of Money during the forties.

Mrs Gore was not afraid to assume the role of lecturer to her readers, setting out the terms of her social investigation in the indispensable prefaces to her works, and then making her presence felt throughout the novels by direct address, moral asides, factual disquisitions, and at times not inconsiderable blasts against Mammon. Such devices are not usually regarded as literary virtues, but they function very well in conjunction with the frequent delights of wit and paradox that enliven the works. Because they are always present as essential ingredients of her style, we are never permitted to lose ourselves fully in either character or 'entourage', as we are always in one sense with

[22] 'Lords and Liveries' (Punch's Prize Novels) in *Punch* 1841; Thackeray: *Works* (1879), xv.

[23] 'Female Novelists—II: Mrs Gore', *New Monthly Magazine*, June 1852, p. 158.

[24] *The Banker's Wife* (1843), iii, Ch. 4.

Mrs Gore in her study, perceiving behind the devices of style a fund of genuine knowledge of the City, and a sober stratum of detached common sense.

Mrs Gore's essentially practical and balanced view of Mammon and all his works is well exemplified in the Preface to her *Men of Capital* (1846), where she writes:

> Few will deny that the age we live in is the age of Money-worship; or that, foremost among the votaries of Mammon, are our own country-people. In Great Britain, however vehement the disputes between High Church and Low, the Molten Calf remains the predominant idol. (i, p. A3.)

This is the familiar argument, a commonplace of so many novels of the period. However, immediately after these words Mrs Gore reveals the other side of the coin. 'That this passion for gold constitutes a fertile source of national greatness, is equally indisputable. But for our appetite for enrichment, our colonies had never been founded, our foreign enterprises never attempted.' (i, p. A3.)

This is the view that both the well-to-do and those aspiring to affluence would have adopted, particularly as it is largely true. The nation grew great financially through the operation of *laissez-faire* principles. Whether it increased in moral stature is another matter, as Mrs Gore proceeds to explain:

> Ambition and Money-love, if they tend to ennoble a country, reduce to insignificance the human particles of which the nation is composed. In their pursuit of riches, the English are gradually losing sight of higher characteristics. (i, p. iv.)

Throughout her works Mrs Gore maintains this balance between genuine recognition of the moral damage occasioned by money-getting, and the commonsense admission of the benefits reaped by the whole nation from commercial expansion. It is this dual approach which makes her the exposer of cant and double standards about money, and which distances her from the assumptions of those authors who allowed themselves all too easily to distort reality for fictional purposes. 'The very sound of a sum in millions', she writes, 'tickles the ears of an Englishman; no man so rich, but endeavours to become richer' (i, p. v). And yet, she avers, it is the false philosopher who regards money with supposed contempt. In *The Moneylender* (1843) she tells her readers:

MONEY is indeed power! Of all the masquerading guises in which false Philosophy loves to parade herself, contempt of MONEY, the ladder by which almost every earthly advantage is attainable, is surely the most absurd![25]

Here Mrs Gore suggests the necessary distinction that is always in her mind between Money and Cupidity. It is the love of money, greed for gain, not money of itself, that constitutes the root of all evils. Love of money transforms the vehicle of 'earthly advantage' into the Mammon of Unrighteousness.

In the figure of Abednego Osalez, the moneylender of the novel's title, Mrs Gore depicts her approach to the problems of a money-centred society by allowing him to create the settings in which he is seen. For purposes of his own that become clear as the novel advances, he chooses these settings quite consciously, and responds to them by assuming attitudes, and even dress, to suit each carefully-created milieu. Underlying the dramatic changes of locale that make Osalez so interesting and unusual is a deliberate moral purpose of his own, which makes the changes credible. Lying behind this interesting technique is Mrs Gore's purpose to use the shifting of scene as a means of modifying the outlook and actions of her principal character in order to convey to the reader her own philosophy.

In the early chapters of the novel Osalez appears as a shabbily-dressed, harsh and vengeful moneylender, known as 'A. O.', who inhabits dingy premises in Greek Street, Soho. The young hero, Basil Annesley, meets him when he is overcome by a bout of illness, and charitably accompanies him home to his mean and wretched house in Paulet Street. Osalez lives up to these squalid surroundings so faithfully that he nearly dies for lack of medical and nursing attention, and Basil shares with the reader a desire to understand what has prompted him to live this kind of life, because, as we very soon learn, Osalez is a scion of a famous European Jewish merchant family, and is enormously rich.

The answer is that Osalez has himself experienced the debilitating effects of Mammon, to which he had turned after personal and social rebuffs in earlier life arising from his racial origin, and is determined, as part of his own redemptive process, to redeem from this curse those members of the aristocracy who borrow from him to vie with each other in acquiring the fripperies of fashion. Behind this, of course,

[25] *The Moneylender* (1843), i, Ch. 4, pp. 92–3.

is Mrs Gore's desire to show that the difference between moneylender and loan-jobber is ultimately one only of degree. The aristocrats who would tolerate the millionaire jobber would despise the humble Jewish moneylender, but, ironically enough, come cringing to him for loans; they must tolerate the meanness and decay around them, as they are the hallmarks of their financial master.

It is in this kind of setting that Osalez delivers his devastating moral rebuke to the Countess of Winterfield:

> 'So long as you enjoy luxuries which you do not and cannot pay for, you are . . . the abject obligee of humble tradesmen. At this moment—woman and Countess as you are—you stand before me as an inferior! Though you may be a Countess of the realm, and I the villified A.O., I rise above you as a capitalist, I rise above you as a moralist . . . (III. Ch. 1, p. 9.)

In marked contrast to this squalid environment with its attendant moral symbolism is Osalez's splendid house in Bernard Street, Russell Square. It is here that he maintains his second identity, which, as was shown in an earlier chapter, is that of international loan-jobber. The house is quietly opulent, and the address a little 'out of the way', so that Osalez's wealth will not be confounded with the ostentation of the West End. Again, the 'entourage' has been consciously chosen by Osalez, and the house in Bernard Street is eminently suitable for entertainment. But it is significant that the guests who assemble there for a sumptuous dinner are not the toadies who flock around Mr Merdle or Mr Melmotte, while secretly despising them. His guests are of his own choosing, and of his own profession. Unlike the Countess and the other titled clients of 'A.O.', who descend into the moneylender's mean den to beg for accommodation, these guests come to Osalez as equals, and are presented as men of almost awesome power. The aristocratic Basil Annesley, who had initially condescended to the old moneylender, now feels himself honoured to be in the presence of this distinguished company.

> For Annesley was beginning to understand with whom he was dipping in the dish. The names by which he heard his companions mutually addressed, were those he knew to be attached to loans and other gigantic financial operations, and saw announced by the papers as having audiences of the Chancellor of the Exchequer . . . (ii, Ch. 4, p. 170.)

In episodes of this kind Mrs Gore emphasizes the value and importance of men of Osalez's class in a way rarely found to any sustained extent in other nineteenth-century novels. Osalez's entourage widens even further when the action of the novel moves to France, where, drawing upon her experiences of that country, where she and her husband had lived in the late thirties, Mrs Gore paints memorable pictures of a society more international in character, in which her protagonist exhibits the geniality and ease that arise from a knowledge that he is both respected and accepted socially. It is in this ambience that Osalez reveals to Basil the story of his life and travels, as the atmosphere makes for intimacy and revelation. Among his journeyings was a pilgrimage to the Holy Land, and his account of this allows Mrs Gore to present her noble condemnation of anti-Semitism, in which she was well ahead of her time.

This, then, is how Mrs Gore solves the problem of character and setting while remaining true to the realities of credit finance and its institutions. At the same time she forges a symbiotic relationship between the young aristocrat and the old jobber which is of mutual benefit to both, and which unifies the varied environments of the novel. The disparate 'entourages' are linked together by Basil's presence in each of them, a presence which arises naturally from the genuine friendship between the two men which transcends the natural prejudices of both. In all the settings described, Osalez enlightens Basil as to the nature of Money as it is, stripped of social cant, and in return the young man rekindles Osalez's suppressed qualities of compassion and leads him to practise a wider charity.

> But the moment a hand so young and stainless as yours poured oil into my wounds, new life was enkindled within me. I seemed to espy noble and undreamed of purposes in MONEY. I began to suspect that it might be converted into a means of human happiness as well as of transitory pleasure. (iii, Ch. 4, p. 137.)

There is a certain satisfaction in seeing the financier fully redeemed from Mammon by the young aristocrat, and the latter made acquainted with the true nature of international and domestic finance. Osalez, like Sidonia, is a brilliant scholar; unlike Sidonia, he can still learn from others.

In *The Moneylender* Mrs Gore is particularly successful in her attempt to liberate the principal character from the tyranny of setting. In the other novels that constitute the group dealing with Mammon the demands are rather different, and more conventional techniques are consequently employed. There is no need to examine the techniques of all these works to note how carefully Mrs Gore uses milieu with functional rather than decorative purposes, and how she never allows selectivity to do violence to the reality of those settings or to the characters who inhabit them. We may conclude with one further example.

In *The Banker's Wife* (1843), written in the same year as *The Moneylender*, the banker Richard Hamlyn is firmly anchored to his opulent London setting, which is meticulously delineated, because it is the physical counterpart of the 'careful and considered deceit' that is the basis of his existence. The 'handsome stablishment in town' is the most vital element of his life, and when he has to leave it, there is nothing left for him but death.

When Hamlyn resides at his country estate of Dean Park, he is made to bring into that ambience the baleful influence of his dishonesty; with his fall, the whole neighbourhood is initially blighted. Nevertheless, Mrs Gore has been quick to indicate, from the opening chapters of the novel, that the snobbery and inward-looking preoccupations of the landed interest in the area blind them to the true nature of the man whom they are despising, not because they suspect him of fraud, but because of their estimate of his social rank. The country estates she describes as

> Affording a favourable type of that rich and smiling order of home landscape which seems almost to embody a portraiture of our social institutions: nothing salient—nothing discordant—a limited horizon—a pleasant foreground, with symbols of peace and prosperity interposing between . . . in order to gratify the taste of those who care less whether Lazarus be sitting famished and suffering at their gates, than that the gates should be of sufficient solidity to exclude the spectacle of so piteous an object. (i, Ch. 1, pp. 3–4.)

This preparation, so early in the novel, of the county setting to which Hamlyn will bring blight and ruin, leaves us in no doubt of Mrs Gore's purpose of including the 'aristocratic' figures of her tale in her scathing condemnation of self-seeking.

It is not likely that Mrs Gore's novels will ever be reprinted, and this is a pity, as few nineteenth-century novelists had her quality of detachment in dealing with commercial matters, or her broadness of vision in involving the whole of society as she observed it in her analysis of Mammon. Of all novelists she seems to be nearest the pulse of society in its real reactions to money and its manifold powers. She sternly castigates the parasite, the usurer, the miser, the man who, through reckless speculation, lusts for gold, and the power that it brings. But at the same time she sees that the revolution in industrial and commercial life that had created the moneyed age had also expanded the country's horizons, and placed it in the foremost rank of European powers. A judicious tribute to this balanced view appeared in Mrs Gore's obituary in the *Illustrated London News*:[26]

> Most women are apt to take the high poetical view of things, and to measure mankind by a constant reference to this standard, so that their heroes and heroines are either angels or devils. Their aspirations are very beautiful, but they are also very deceptive; and Mrs Gore avoided them in order to teach the homespun useful lesson of contentment.

[26] *Illustrated London News*, 16 Feb. 1861, p. 147.

BIBLIOGRAPHY

1. Reference Works

(a) General

Dictionary of National Biography, The (1922 edn.)
Elton, O., *A Survey of English Literature 1830–1880* (2 vols., 1927)
Kunitz, S. J. and Haycraft, H., *British Authors of the Nineteenth Century* (New York, 1936)
New Cambridge Bibliography of English Literature, The, vol. iii (1969)

(b) Historical and Special

Art Journal Catalogue of the Great Exhibition, The (1851)
Jerrold, B., *London, a Pilgrimage* (1872)
Macaulay, T. B., *History of England* (1855)
Quennell, P., *London's Underworld* (1950)
Turberville, A. S., *The Spanish Inquisition* (1932)
Young, G. M. (ed.), *Early Victorian England* (2 vols., 1934)

2. Works on Economic Theory and History

(a) Eighteenth Century and earlier

Gibbon, E., *Miscellaneous Works*, vol. I. (1796)
Grotius, H., *De Jure Belli ac Pacis* (1625)
Hume, D., *Essays Moral, Political and Literary* (1741–2) Ed. T. H. Green and T. H. Grose (1875)
Malthus, T., *Essay on Population* (1798)
Smith, A., *Inquiry into the Nature and Causes of the Wealth of Nations* (1776)
—— *Theory of the Moral Sentiments* (1759)

(b) Nineteenth Century

Jevons, W. S., *Theory of Political Economy* (1871)
Marshall, A., *Principles of Economics* (1890)
Mill, James, *Elements of Political Economy* (1821)
—— 'The Formation of Opinions', in *The Westminster Review* (1826), vi.
Mill, J. S., *Principles of Political Economy* (1848)
Price, L. L., *A Short History of Political Economy in England* (3rd edn. 1900)
Ricardo, D., *Principles of Political Economy and Taxation* (1817) in *Works and Correspondence*, ed. P. Sraffa and M. H. Dobb (11 vols., Cambridge 1951–7), vol. I.

Smart, W., *Introduction to the Theory of Value* (1891)
Stephen, L., *The English Utilitarians* (1900)
Tomkins, I., *Thoughts upon the Aristocracy of England* (11th edn. 1835)

(c) Twentieth Century

Clapham, J. H., *An Economic History of Modern Britain* (2 vols., 1926, 2nd edn. 1930)
Johnson, E. A. J., *Predecessors of Adam Smith* (1937)
Melville, L., *The South Sea Bubble* (1921)
Perkin, H., *The Origins of Modern English Society 1780–1880* (1969)
Rees, J. F., *A Short Fiscal and Financial History of England 1815–1918* (1921)

3. Works on Commercial and Mercantile History

Clapham, J. H., 'Life in the New Towns', in G. M. Young (ed.), *Early Victorian England* (2 vols., 1934, vol. 1, pp. 225–44)
Clapham, J. H., 'Work and Wages', in G. M. Young, op. cit., vol. i, pp. 1–76.
Fox Bourne, H. R., *English Merchants* (1866), New edn. 1886
—— *The Romance of Trade* (n.d.)
Levi, L., *History of British Commerce* (1st edn. 1872; 2nd edn. 1880)
Morier Evans, D., *The Commercial Crisis 1847–1848* (1849)
—— *Facts, Failures and Frauds* (1859)
—— *The History of the Commercial Crisis 1857–1858* (1859)
Mottram, R. H., 'Town Life', in G. M. Young, op. cit., vol. 1, pp. 179 f.
Penson, L. M., 'The London West India Interest in the Eighteenth Century', in *English Historical Review* (July 1921, pp. 373–92)
Potter, E., *A Picture of a Manufacturing District* (Manchester 1856)
Rogers, J. E. T., *Industrial and Commercial History of England* (1892)
Smart, W., *Economic Annals of the Nineteenth Century* (2 vols. 1910)
Stokes, M. V., 'Charles Dickens: a Customer of Coutts & Co.', in *The Dickensian* (Jan. 1972)
Tooke, T., *History of Prices* (1838–1857)
Wade, J., *British History, Chronologically Arranged* (1839; many revisions and editions subsequently)

4. Specialized Works on Commercial Institutions

(a) Banking

Andréadès, A., *History of the Bank of England 1640–1903* (C. Meredith, tr., 2 vols. in 1, 1909.)
Ashby, J. F., *The Story of the Banks* (1934)
Clapham, J. H., *The Bank of England, a History 1694–1914* (2 vols., Cambridge 1944)
Francis, J. A., *History of the Bank of England* (2 vols., 3rd edn. 1848)

Pressnell, L. S., *Country Banking in the Industrial Revolution* (Oxford 1956)
Sykes, J., *The Amalgamation Movement in English Banking 1825–1934* (1936)

(b) The Stock Exchange

Duguid, C., *The Story of the Stock Exchange* (1910)
Mortimer, T., *Every Man His Own Broker* (1761)
—— *The Nefarious Practice of Stockjobbing Unveiled* (1810)

(c) Insurance

Francis, J. F., *Annals, Anecdotes and Legends of Life Assurance* (1853)
Walford, C., *The Insurance Cyclopaedia* (5 vols. 1871–8)
Withers, H., *Pioneers of British Life Assurance* (1951)

(d) Miscellaneous Studies

Darwin, B., 'Country Society', in G. M. Young, op. cit., vol. I pp. 262 f.
Disraeli, B., *An Inquiry into the Plans, Progress and Policy of American Mining Companies* (1825)
—— *Lawyers and Legislators* (1825)
English, H., *A Complete View of the Joint Stock Companies . . .* (1827)
Francis, J. A., *A History of the English Railway* (2 Vols. 1851)
Lewin, H. G., *The Railway Mania and its Aftermath* (1936)
McCulloch, J. R., *A Dictionary of Commerce* (1832) (Refs. in text are to the edition of 1845)
—— *Statistical Account of the British Empire* (1837)
Newmarch, W., 'On the Recent History of the Crédit Mobilier', in *Journal of the Statistical Society*, Series A, vol. xxi, Dec. 1858, pp. 444 ff.
Wordsworth, C. C. F., *Law of Joint Stock Companies* (1842)

5. Biographical Works on Noted Capitalists and Men of Business

Coleridge, E. H., *The Life of Thomas Coutts, Banker* (2 vols., 1920)
Corti, Egon Caesar Graf von, *Der Aufsteig des Hauses Rothschild* (Leipzig 1927), trans. B. and B. Lunn as *The Rise of the House of Rothschild* (New York 1928)
Cowles, V., *The Rothschilds, a Family of Fortune* (1973)
Rothschild, The Lord, *The Shadow of a Great Man* (London: privately printed, 1982)
Carmyllie, R. R., *Charles Dickens and the 'Cheeryble' Grants* (Ramsbottom: privately printed, 1981)
Elliot, W. H., *The Story of the 'Cheeryble' Grants* (Manchester 1906)
Knapp, A. and Baldwin, W., 'Henry Fauntleroy, Executed for Forgery', in *The New Newgate Calendar* (*c.*1826); Folio Society, 1960
Hudson, George, 'Mr George Hudson and the English Railway System', in *The Bankers' Magazine* Dec. 1851, pp. 746–54

——'Mr. George Hudson', in *Fraser's Magazine* Aug. 1847, pp. 215–22
Puckler-Muskau, Prince, *Tour of a German Prince* (London, 1832)

6. Novels, Poems and Plays

These are listed alphabetically by author. The date of publication given is that of the first edition of the work in book form. Earlier publishing history, where relevant, is given in the text and in the Notes.

Anon, *Yesterday: a Novel* (1859)
Aytoun, W. E., *Norman Sinclair* (1861)
Bell, R., *The Ladder of Gold* (1850)
Bulwer, E. (Lord Lytton), *The Caxtons* (1849)
—— *The Disowned* (1828)
—— *Eugene Aram* (1832)
Byron, G. G., Lord, *Don Juan* (1822)
Byron, H. J., *Paid in Full* (1865)
Chichester, F. R., *Masters and Workmen* (1851)
Cook, E. D., *The Trials of the Tredgolds* (1864)
Costello, D., *The Millionaire of Mincing Lane* (1857)
Dickens, C., *Bleak House* (1853)
—— *The Chimes* (1845)
—— *Christmas Books* (Penguin English Library, 2 vols., ed. Michael Slater, 1971)
—— *A Christmas Carol* (1843)
—— *Dombey and Son* (1848)
—— *Hard Times* (1854)
—— *Hunted Down* (1859)
—— *Little Dorrit* (1857)
—— *Martin Chuzzlewit* (1844)
—— 'The Mudfog Papers' (in *Bentley's Miscellany*, 1837–8)
—— *Nicholas Nickleby* (1839)
—— *The Pickwick Papers* (1837)
—— *A Tale of Two Cities* (1859)
—— *The Uncommercial Traveller* (1861)
Disraeli, B., *Coningsby* (1844)
—— *Sybil* (1845)
—— *Tancred* (1847)
—— *The Voyage of Captain Popanilla* (1828)
Edgeworth, M., *Harrington* (1817)
'Eliot, George' (Mrs. J. W. Cross) *Middlemarch* (1872)
Fonblanque, A., *A Tangled Skein* (1862)
Fullom, S. W., *The Great Highway* (1854)
Galsworthy, J., *The Skin Game* (1920)
Gaskell, Mrs E., *North and South* (1855)

Gore, Mrs C., *Cecil, or the Adventures of a Coxcomb* (1841)
—— *The Banker's Wife* (1843)
—— *Men of Capital* (1846)
—— *The Moneylender* (1843)
—— *Progress and Prejudice* (1854)
Harwood, J. B., *Lord Ulswater* (1867)
Higgins, M. J., 'Jacob Omnium, the Merchant Prince', in *New Monthly Magazine*, Aug. 1845, pp. 567–78
Kinglake, A., *Eothen* (1844)
Kingsley, C., *Yeast* (1851)
Lever, C., *The Bramleighs of Bishop's Folly* (1868)
—— *That Boy of Norcott's* (1869)
Martineau, H. 'Berkeley the Banker' (*Illustrations of Political Economy* No. 14, 1833)
Moore, T., *Lalla Rookh* (1817)
Mulock, D. M. (Mrs Craik) *John Halifax, Gentleman* (1856)
Pardoe, J., *The Poor Relation* (1858)
Peacock, T. L., *Crotchet Castle* (1831)
Reade, C., *Hard Cash* (1863)
Robinson, E., *The Gold Worshippers* (1851)
Shadwell, T., *The Volunteers: or The Stockjobbers* (1693)
Sinclair, C., *Sir Edward Graham, or The Railway Speculators* (1849): reissued as *The Mysterious Marriage: or Sir Edward Graham* (1854)
Smedley, F., *Frank Fairlegh* (1850)
Swift, J., *The Dog and Thief* (1726)
—— *Pastoral Dialogue: Marble Hill* (1727)
—— *The South Sea Project* (1721)
Thackeray, W. M., 'Codlingsby' (in *Punch* 1847; also in *Works*, 1879, vol. 15)
—— *Diary of C. Jeames de la Pluche, Esq., With His Letters* (1845–46), *Works* (1879), vol. 15
—— *History of Samuel Titmarsh and the Great Hoggarty Diamond* (1849)
Trollope, A., *Barchester Towers* (1857)
—— *The Way We Live Now* (1875)
Warren, S., *Passages from the Diary of a Late Physician* (1838)
Wills, W. G., *Notice to Quit* (1861)

7. Biography, Essays, and Belles Lettres

Baring-Gould, S., *Early Reminiscences* (1923)
Blake, R., *Disraeli* (1966)
Carlyle, T., *Latter-Day Pamphlets* (1850)
—— *Past and Present* (1843)
—— 'Signs of the Times', in *Edinburgh Review*, xlix, June 1829
Cruikshank, G., *Comic Almanack for 1835* (1835)
—— *Table Book for 1845* (1845)

Escott, T. H. S., *Anthony Trollope* (1913)
Forster, J., *Life of Charles Dickens* (7th edn., 3 vols., 1872)
Fox, C., *Memories of Old Friends 1835–1871* (New and Revised edn., ed. H. N. Pym, 1883)
Greville, C. C. F., *The Greville Memoirs* (ed. H. Reeve, 8 vols., 1913)
Head, Sir Francis B., *Rough Notes* (1826)
Kohl, J. G., *England and Wales* (1844)
Meynell, W., *Benjamin Disraeli, an Unconventional Biography* (1903)
Mill, J. S., *Autobiography* (2nd edn. 1873)
Pilgrim Edition of the Letters of Charles Dickens (ed. M. House and G. Storey), 1965 continuing.
Pope-Hennessey, U., *Charles Dickens* (1945; Reprint Soc. edn. 1947)
Trollope, A., *An Autobiography* (1883), World's Classics edn. 1968
Wellington, Duke of, *Dispatches . . . 1799–1818* (Gurwood's Edition, 12 vols., 1837–38)

8. Works of Literary Criticism and Special Studies

Altick, R. D., *Victorian People and Ideas* (1973)
Athill, R., 'Dickens and the Railway', in *English,* No. 13 (1961), pp. 130–5
Aydelotte, W. O., 'The England of Marx and Mill as reflected in Fiction', in *Journal of Economic History* (Supplement: 'Tasks of Economic History'), No. 88, 1948)
Bejamin, E. B., 'The Structure of *Martin Chuzzlewit*', in *Philological Quarterly*, No. 34 (1955), pp. 39–47
Booth, B. A., 'Trollope and *Little Dorrit*', in *The Trollopian* No. 2 (1948), pp. 237–40
Brightfield, M. F., 'The Coming of the Railroad to Early Victorian England as viewed in the Novels of the Period 1840–1870', in *Technology and Culture* No. 3 (1962), pp. 45–72
Brogunier, J., 'The Dreams of Montague Tigg and Jonas Chuzzlewit', in *The Dickensian* No. 58 (1962), pp. 165–70
Brown, J. M., *Dickens: Novelist in the Market-Place* (1982)
Butt, J., 'The Topicality of *Little Dorrit*', in *University of Toronto Quarterly* No. 29 (1959), pp. 1–10
—— and Tillotson, K., *Dickens at Work* (1957)
Cazamian, L., *Le Roman social en Angleterre* (1903) (Tr. M. Fido, *The Social Novel in England*, 1973)
Chaudhury, G. A., 'The Mudfog Papers', in *The Dickensian*, May 1974
Cline, C. L., 'Disraeli and Thackeray', in *Review of English Studies* vol. 19, No. 76 (1943), p. 406
Cockshut, A. O. J., *Anthony Trollope: a Critical Study* (1955)
Dabney, R. H., *Love and Property in the Novels of Dickens* (1967)
Dodds, J. W., *The Age of Paradox* (1953)

Elwin, M., *Thackeray: a Personality* (1932)
Fielding, K. J., *Charles Dickens, a Critical Introduction* (1958)
Frietzsche, A. H., 'Action is Not for Me', in Utah Academy of Sciences, *Proceedings* (1959–60), pp. 45–9
(Gore, Mrs C.) 'Female Novelists II: Mrs. Gore', in *The New Monthly Magazine*, June 1852, pp. 157–168
—— Obituary: *Gentleman's Magazine*, Mar. 1861, pp. 345–6
—— Obituary: *Illustrated London News*, Feb. 1861, p. 147
Gregg, W. R., Review of *Alton Locke* in *Edinburgh Review* vol. xciii, (1851)
Hardy, B., *The Moral Art of Dickens* (1970)
Herring, P. D., 'Dickens' Monthly Number Plans for *Little Dorrit*', in *Modern Philology*, Aug. 1966, pp. 23–63
House, H., *The Dickens World* (1941)
Jackson, H., *Dreamers of Dreams* (1948)
Jansonius, H., 'Some Aspects of Business Life in Early Victorian Fiction' (Doctoral thesis, Amsterdam 1926)
(Lewes, G. H.)., Review of *Coningsby* in *British Quarterly Review* No. 10 (1849), pp. 120 f.
Lewis, C. J., 'Disraeli's Concept of Divine Order', in *Jewish Social Studies* No. 24 (1962), pp. 145 f.
Levine, R. A., 'Disraeli's *Tancred* and "The Great Asian Mystery"', in *Nineteenth Century Fiction*, No. 22 (1968), pp. 71–85
Lowell, J. R., Review of *Tancred* in *North American Review* No. 65 (1847), pp. 83 f.
Milnes, R. M., Review of *Tancred* in *Edinburgh Review* No. 86 (July 1847), pp. 138–55
Rosenberg, E., *From Shylock to Svengali* (Stanford 1961)
Russell, N., '*Nicholas Nickleby* and the Commercial Crisis of 1825', in *The Dickensian* 1981, pp. 144–50
Sadleir, M., *Trollope: a Commentary* (1927)
Smith, Grahame, *Dickens, Money and Society* (1968)
——, Grahame and Angela, 'Dickens as a Popular Artist', in *The Dickensian*, Sept. 1971, pp. 131 f.
Steig, M., '*Dombey and Son* and the Railway Panic of 1845', in *The Dickensian* (1971), pp. 145–48
Thompson, D. G., *The Philosophy of Fiction in Literature* (New York 1890)
Tillotson, K., *Novels of the Eighteen Forties* (1961)

Index